# Affect and Mathematical Problem Solving

Douglas B. McLeod   Verna M. Adams
Editors

# Affect and Mathematical Problem Solving

## A New Perspective

With 16 Illustrations

Springer-Verlag New York Berlin Heidelberg
London Paris Tokyo

Douglas B. McLeod
Elementary and Secondary Education
Washington State University
Pullman, Washington 99164-2122, USA

Verna M. Adams
105 Aderhold Hall
University of Georgia
Athens, Georgia 30602, USA

Library of Congress Cataloging-in-Publication Data
Affect and mathematical problem solving : a new perspective / Douglas
  B. McLeod, Verna M. Adams, editors.
     p.   cm.
  Includes bibliographies and index.
  ISBN 0-387-96924-1 (alk. paper)
  1. Problem solving—Study and teaching—Psychological aspects.
  2. Mathematics—Study and teaching—Psychological aspects.
  3. Affect (Psychology)  I. McLeod, Douglas B.  II. Adams, Verna M.
  QA63.A34     1989
  153.4′3—dc 19                                          88-39063

Printed on acid-free paper.

Typeset by Publishers Service, Bozeman, Montana.
Printed and bound by Edwards Brothers, Ann Arbor, Michigan.
Printed in the United States of America.

9 8 7 6 5 4 3 2 1

ISBN 0-387-96924-1 Springer-Verlag New York Berlin Heidelberg
ISBN 3-540-96924-1 Springer-Verlag Berlin Heidelberg New York

# Preface

Research on mathematical problem solving has received considerable attention in recent years, not only from those who do research on mathematics learning and teaching, but also from psychologists and others who work in the cognitive sciences. Although most of this research has focused on cognitive factors, there has been a surge of interest in the role of affect in mathematical problem solving. This book explores these affective factors and their relationships to the cognitive processes involved in problem solving. The ideas are of particular interest to those who work in mathematics education, psychology, and other fields that are related to the cognitive sciences.

The origins of the book can be found in George Mandler's (1984) theory of emotion and in Edward Silver's (1985) effort to integrate the multiple perspectives that influence research on mathematical problem solving. Therefore, the interdisciplinary mix for this volume includes not only the usual perspectives of mathematics education, cognitive psychology, and artificial intelligence, but also cognitive approaches to clinical and counseling psychology.

The book is divided into five parts. The first introduces the basic ideas of Mandler's theory and suggests how the theory can be applied in research on mathematical problem solving. Part II presents five studies of learning that test and extend the theory, and the five chapters in Part III perform the same function in the context of research on teaching. Part IV includes two responses to the theory, one from the viewpoint of mathematics education, and the other from cognitive psychology. Finally, following George Polya's recommendations, Part V is spent "looking back" at the problem of how to build a theory of affective factors in mathematical problem solving.

The development of an interdisciplinary volume like this one requires a substantial amount of time both for interaction and reflection by the participants. In this case we were fortunate to have the support of the National Science Foundation for two meetings where the chapter authors and other researchers had a chance to debate the issues and report on their own research. The first meeting was held in May 1986, in San Diego, and focused on presenting and discussing Mandler's theory. Participants also outlined their plans for implementing Mandler's ideas in their own research during the next year. In June 1987, the participants met again in San Diego and reported on the results of their efforts to incorporate affective factors in their research on mathematical problem solving. This volume grew out of the papers that were presented at the second meeting. In addition to these two meetings, subgroups of participants presented symposia at the 1987 Annual Meetings of both the American Educational Research Association and the National Council of Teachers of Mathematics. Consequently, we have here not the typical report of the proceedings of a conference, but the results of two years of study, debate, and research by the participants.

The preparation of this volume was supported in part by National Science Foundation Grant No. MDR-8696142, as well as by other grants from several agencies to some of the individual authors. Any opinions, conclusions, or recommendations are those of the authors and do not necessarily reflect the views of the funding agencies. We also want to acknowledge the cooperation of both of the academic institutions that have served as the host for this project, San Diego State University and Washington State University.

We want to thank all of the authors for their contributions and cooperation. We are especially grateful to George Mandler of the University of California at San Diego for his willingness to share his psychological theories and to see them tested in the domain of mathematics classrooms. Special thanks are also due to Elizabeth Fennema of the University of Wisconsin, who shared generously of her experience in research on affect and mathematics education. Fennema and Mandler both served on our interdisciplinary Project Advisory Board, along with David Carlson, a research mathematician at San Diego State University, and F.J. Crosswhite, Past-President of the National Council of Teachers of Mathematics. The Advisory Board, along with Ray Hannapel of the National Science Foundation, played an important role in the development of the project, and we are grateful for their help.

In addition, we want to thank the staff members in San Diego and Pullman who have assisted us with this project, especially Kathie Duncan. We also want to thank the people at Springer-Verlag for their help in producing this book. Finally, we thank our families for their support throughout this project.

Pullman, Washington                                          Douglas B. McLeod
Athens, Georgia                                                Verna M. Adams

# Contents

# Contributors

VERNA M. ADAMS
Department of Mathematics Education, University of Georgia, Athens, Georgia 30602, USA

PAUL COBB
Department of Education, Purdue University, West Lafayette, Indiana 47907, USA

KATHLEEN CRAMER
Department of Elementary Education, University of Wisconsin-River Falls, River Falls, Wisconsin 54022, USA

ELIZABETH FENNEMA
Wisconsin Center for Education Research, University of Wisconsin-Madison, Madison, Wisconsin 53706, USA

JOE GAROFALO
Department of Curriculum Instruction and Special Education, University of Virginia, Charlottesville, Virginia 22903, USA

DOUGLAS A. GROUWS
Department of Curriculum and Instruction; Center for Research in Social Behavior, University of Missouri, Columbia, Missouri 65211, USA

LAURIE E. HART
Department of Elementary Education, University of Georgia, Athens, Georgia 30602, USA

JAMES J. KAPUT
Department of Mathematics, Southeastern Massachusetts University, North Dartmouth, Massachusetts 02747; Educational Technology Center, Harvard University, Cambridge, Massachusetts 02138, USA

DIANA LAMBDIN KROLL
School of Education, Indiana University, Bloomington, Indiana 47405, USA

FRANK K. LESTER
School of Education, Indiana University, Bloomington, Indiana 47405, USA

GEORGE MANDLER
Center for Human Information Processing, University of California–San Diego, La Jolla, California 92093, USA

SANDRA P. MARSHALL
Department of Psychology; Center for Research in Mathematics and Science Education, San Diego State University, San Diego, California 92182, USA

BARBARA A. MCDONALD
Navy Personnel Research and Development Center, San Diego, California 92107, USA

DOUGLAS B. MCLEOD
Elementary and Secondary Education, Washington State University, Pullman, Washington 99164-2122; Department of Mathematical Sciences, San Diego State University, San Diego, California 92182, USA

WENDY METZGER
Department of Mathematics, Palomar College, San Marcos, California 92069, USA

EDWARD A. SILVER
School of Education; Learning Research and Development Center, University of Pittsburgh, Pittsburgh, Pennsylvania 15260, USA

JUDITH THREADGILL SOWDER
Department of Mathematical Sciences, San Diego State University, San Diego, California 92182, USA

LARRY SOWDER
Department of Mathematical Sciences, San Diego State University, San Diego, California 92182, USA

ALBA G. THOMPSON
Department of Mathematics, Illinois State University, Normal, Illinois 61761, USA

PATRICK W. THOMPSON
Department of Mathematics, Illinois State University, Normal, Illinois 61761, USA

TERRY WOOD
Department of Education, Purdue University, West Lafayette, Indiana 47907, USA

ERNA YACKEL
Department of Mathematical Sciences, Purdue University Calumet, Calumet, Indiana 46323, USA

# Part I A Theory of Affect for Mathematical Problem Solving

# 1
# Affect and Learning: Causes and Consequences of Emotional Interactions

GEORGE MANDLER

Affect is the least investigated aspect of human problem solving, yet it is probably the aspect most often mentioned as deserving further investigation. The "problem-solving" and "teaching-and-learning" literature is full of remarks that have a single message: "Someday soon—maybe tomorrow—we must get around to doing something about affect and emotion." I am delighted to see that "tomorrow" has come.

In a sense, however, tomorrow came some time ago, and I have been puzzled by the fact that an extensive literature on stress and cognitive efficiency rarely, if ever, crossed the disciplinary fences and reservations. I refer here to the continuing interest in the way motives, stress, emotions, and affects interact with the acquisition of skills, and the storage and recovery of memories. That particular tradition goes back to the beginning of the century with the Yerkes-Dodson law, which introduced the notion that the relation between stress and efficiency of learning was curvilinear—worst at both weak and strong levels of stress and best at intermediate levels (Yerkes & Dodson, 1908). In modern times, it is represented by Easterbrook's influential paper (1959), in which he hypothesized that our attentional capacity for picking up information is increasingly impaired by increases in emotional intensity. Major treatments within the information-processing context were provided by Broadbent's *Decision and Stress* (1971) and Kahneman's *Attention and Effort* (1973). I have discussed these various approaches in the context of my own view of stress and efficiency in my book *Mind and Body* (1984), and I develop this theme further within appropriate contexts to follow.

First, I present a brief outline of my constructivist view of emotion.[1] This outline is followed by a discussion of some possible applications of those notions to problem solving and learning. A microanalytic approach is stressed, and questions are asked about the uses of affect and the specific effect of human error.

Before I proceed, a brief word about the word *affect*. Unfortunately, the term has meant many things to many people, acquiring interpretations that range from "hot" to "cold." At the hot end, affect is used coextensively with the word *emotion*, implying an intensity dimension; at the cold end, it is often used without passion, referring to preferences, likes and dislikes, and choices. I interpret the use that it has acquired in the problem-solving field to be hot, rather than cold, and I frequently use the words *affect* and *emotion* interchangeably in this chapter.

## The Construction of Emotion

Much contemporary cognitive theory would lead one to believe that human beings are unable to feel. Until recently, conventional wisdom in cognitive science described people as passionless creatures who think and act rationally and coolly. Such an approach has been most obvious in the investigation of problem solving. In reality, psychologists and other human beings typically are frustrated, angry, joyous, delighted, intense, anxious, elated, and even fearful when dealing with complex problems, such as a recalcitrant computer program, an exasperated customer, a difficult bridge hand, an unassembled toy, or a mathematics problem. If cognitive psychology aspires to an understanding of human thought and action, it can ill afford to leave out their emotional aspects. I should add that it is not enough for cognitive models to have nodes or processors labeled *fear* or *joy* that can be accessed whenever appropriate. That approach simply acknowledges theoretical impoverishment; it does not solve the problem of how emotions arise or how they are to be represented.

Emotion is not only anecdotally and phenomenally part of human thought and action; there is now a burgeoning body of evidence that emotional states interact in important ways with traditional cognitive functions. For example, Isen's work has shown that positive feelings determine the accessibility of mental contents in the process of decision making, serve as retrieval cues, and influence problem-solving strategies (e.g., Isen, Means, Patrick, & Nowicki, 1982). More generally, accessibility of mental contents is determined by the mood both at the time of original encounter and at the time of retrieval (Bower, 1981; Bower & Mayer, 1985). Thus, emotion should be of both general and specific interest to cognitive psychologists.

There are essentially two different views of emotional phenomena. One view is that emotions are discrete patterns of behavior, experience, and neural activity. These patterned packages are considered to be fundamental emotions (such as fear, joy, and anger), and are usually restricted to some 5 or 10 in number. I shall refer to this as the fundamentalist position, with Charles Darwin as its patron saint. It is somewhat surprising that after more than 100 years of searching, the fundamental emotion theorists cannot agree on what these *basic* emotions are. The basic emotions usually are considered to have developed as a function of human (and mammalian) evolution, with other emotions being some combination of the fundamental ones. Given the wide variety of human emotion, this latter postulate engenders complex analyses of and recipes for emotion—something like a cordon bleu school of emotion.

The second approach is cognitive and constructivist, which considers emotional experience (and behavior) to be the result of cognitive analyses and physiological (autonomic nervous system) response (Averill, 1980; Lazarus, Kanner,

& Folkman, 1980; Mandler, 1984). The founding father of the constructivist approach is William James, but his particular constructions depended entirely on patterns of visceral and muscular feedback and are no longer considered acceptable.

Another dichotomous argument in the field of emotion concerns the character of emotional or affective reactions. This argument is related to, but somewhat different from, the division between fundamentalist and constructivist theories. The distinction here is between a constructivist view of emotional reactions (specifically subjective feelings) and one that makes a distinction between affective and cognitive analyses. The foremost practitioner of the latter approach is Robert Zajonc. He has marshalled an impressive array of anecdotal and phenomenal evidence to argue that affective responses are unmediated and are fast initial reactions to people and events (Zajonc, 1980). Affective reactions are said to occur (without mediating cognitive analyses) in response to specific aspects of an event. The implication is that specific characteristics of events (so-called "preferanda") force preferences. The response of a cognitive constructivist to this view is that the attributes and features in the environment need to be analyzed and processed by underlying representations. Specifically, the phenomenal evidence that many of our reactions to the world are initially affective is not, by itself, evidence for an unmediated affective response. Affective experiences are constructed conscious contents, just as any other such contents. The indisputable observation that we frequently react affectively to events, before experiencing a more "analytic" knowledge of the event, speaks to the ubiquity of affective and evaluational constructions and intentions. We live in a world of value and affect, and the themes that determine our conscious constructions often require an affective content. This does not force an absence of other analyses and activations going on at the same time at the preconscious level. The analyses to be used in conscious constructions will be determined by the intentions and requirements of the moment, which, in many cases, happen to be affective. On the other hand, the assertion of "fast" affective reactions is an empirical one. If, in fact, affective, evaluative reactions are faster than those that require only access to cognitions (knowledge), then one would have to reconsider the constructivist argument. However, the available evidence suggests that affective reactions are actually *slower* than cognitive ones (e.g., Mandler & Shebo, 1983). Such a finding is consistent with the notion that questions about the familiarity or identity of objects or events require access to relatively fewer features of the underlying representation and can be quickly constructed; by comparison, questions about the appeal, beauty, or desirability of the object or event require more extensive analyses and constructions.

The construction of emotion, in general – at least as seen by this practitioner – consists of the concatenation in consciousness of some cognitive evaluative schema with the perception of visceral arousal. We experience unitary emotions, often labeled with varying consistency with such names as *fear, anger, anxiety,* and *jealousy*. Only rarely do we experience more than one emotion at one time. These conscious constructions are, as are all other such constructions, holistic unitary experiences, even though they may derive from separate and even independent schematic representations. This view of the construction of conscious experience suggests that the conscious contents of the moment are constructed out of two or more underlying structures and have the function of "making sense" of the current situation (Mandler & Nakamura, 1987). This attitude toward emotion only approximates the common-sense meaning of the term. The question "What is an emotion?" is not, in principle, answerable. The term is a natural language expression that has all the advantages (communicative and inclusive) and disadvantages (imprecise and vague) of the common language. It is precisely for communicative purposes, however, that one needs to approximate the common meaning as a first step.

Of the possible analyses of common language emotions, I have focused on two characteristics: the notion that emotions express some aspect of value, and the assertion that emotions are hot, implying a gut reaction or a visceral response. These two aspects not only speak to the common usage, but they also reflect the fact that conscious constructions are frequent occurrences in everyday life; thus, an analysis of the concatenation of value and visceral arousal addresses both natural language usage and a theoretically important problem. One of the consequences of such a position is that it leads to the postulation of a potentially innumerable number of different emotional states, because no situational evaluation will be the same from one occasion to the next. There are, of course, regularities in human thought and action that produce general categories of these constructions—categories that have family resemblances and overlap in the features that are selected for analysis and that create the representation of value (whether it is the simple dichotomy of good and bad, the appreciation of beauty, or the perception of evil). These families of occasions and meanings construct the categories of emotions found in the natural language (and psychology). The commonalities found within these categories may vary from case to case, and they have different bases for their occurrence. These commonalities may be based on the similarity of external conditions, as in the case of some fears and environmental threats. Sometimes an emotional category may be based on a collection of similar behaviors, such as the subjective feelings of fear related to flight. The common category may arise from a class of incipient behaviors, as in hostility and destructive action. Hormonal and physiological reactions may provide a common basis for the experience of lust. Purely cognitive evaluations may generate judgments of helplessness that eventuate in anxiety. These commonalities give rise to the appearance of fundamental or discrete emotions. With respect to the neural aspect of discrete emotions, I do not believe that we have heard the last word on the assertion that certain emotions are identified with specific localized brain functions. The function of the limbic system, for example, in the generation of emotional experience is vague, but the elicitation of specific actions (e.g., "rage" in cats) can be seen as just that—actions. We do not know whether such elicited actions produce any accompanying subjective emotional experiences. I would argue initially that such actions may be constituents in the construction of subjective emotions.

Emotions are usually situation specific, and subjective emotional states, however one defines their source, need to be tied to cognitive evaluations that "select" the appropriate emotion. The various indices that constitute an emotion are not haphazard collections of current conditions; they are organized by behavioral, cognitive, or physiological states and conditions of the individual. The source for the discreteness of the emotions can often be found in those conditions. One should understand, however, that even these "fundamental" emotional states require some analytic, cognitive processing. Even if it were the case that well-coordinated neural/behavioral/cognitive systems are "elicited" as a unit, as single emotions, one would still have to analyze the eliciting conditions that produce one emotion or another.

The problem of cognitive evaluation seems to be common to all emotion theories. In recent years, there has been an active search for the basis of these valuative structures. Basically, cognitive evaluations require a theory for the representation of value. That task cannot be avoided if we are truly interested in the full range of human thought and action. What is the mental representation that gives rise to judgments and feelings of good or bad or of some affective nature in general? Surprisingly, psychologists have paid relatively little attention to problems of value, in the sense of developing a theory of the underlying structures that give rise to phenomenal experiences of value. My concern is with both accepting and explaining these phenomenal experiences.

A full discussion of the problem of value is precluded in this presentation. In brief it involves the different external and internal sources that lead us to see some person or event as good or bad, as evil or benign, as harmful or beneficial. Within the realm of learning and problem solving, there is a wide range of sources of values. On the one hand, it is considered a good thing to be proficient in mathematics; on the other hand, mathematics is a source of apprehensions. The well-known phenomenon of *number shock* demonstrates an underlying value that mathematical manipulation is difficult, complex, and potentially frightening. Many people in our society confess helplessness when dealing with numbers; others admire those who perform even arithmetic without pain. Also relevant to the problem-solving context is the value inherent in the completion of a problem or a task. The very fact that a task is done, often regardless of whether it is completed successfully, provides a source of positive value. All of these values will feed into the construction of emotions during mathematical problem solving. More specific is the subjective evaluation of an error, a mistake, or a success in the process of cognitive processing,[2] which will be addressed in more detail later in this chapter.

If evaluative cognitions provide the quality of an emotional experience, then visceral activity provides its intensity and peculiar emotional feel. Some theorists, and some critics of my position, have argued that the autonomic, visceral component is not necessary for emotional experience. In reply, I would argue that emotion is, of course, one of those great natural language concepts that can serve any and most arguments, depending on how one wishes to use the language or the concept. The best we can do is to propose a definition that satisfies a reasonable portion of the common concept and produces some degree of social consensus. In part, the denial of visceral activity as a necessary part of "real" emotions is related to such alternative terms as *affect*. One may say that something is pretty or fine or awful or even disgusting quite dispassionately and unemotionally. I believe that we need to understand the occasions when visceral activity, however slight, co-occurs with these judgments or affects.

In one version of the common understanding of emotion, the occurrence of some visceral or gut reaction is generally assumed. Emotions are said to occur when we feel aroused, agitated, when our "guts are in an uproar," and so forth. Reference is always made (and properly so) to some autonomic nervous system activity, such as increased heart rate, sweating, or gastrointestinal upheavals. The autonomic nervous system (ANS) has been implicated in quasi-emotional activity ever since Walter Cannon delineated the function of the sympathetic and parasympathetic systems in fight/flight reactions, giving them a role beyond that of expending and conserving energy in keeping the internal environment stable. If one looks at the literature on the ANS, however, one is faced with a lack of any principled account of the sources of ANS activation.

I have argued that on a majority of occasions, visceral arousal follows the occurrence of some perceptual or cognitive discrepancy or the interruption or blocking of some ongoing action. Such discrepancies and interruptions depend to a large extent on the organization of mental representation of thought and action. Within the purview of schema theory, these discrepancies occur when the expectations of some schema are violated. This is the case whether the violating event is worse or better than expected and accounts for visceral arousal in both unhappy and joyful occasions. Most emotions follow such discrepancies because the discrepancy produces visceral arousal. The combination of that arousal with an ongoing evaluative cognition produces the subjective experience of an emotion. I do *not* say that emotions are interruptions. Interruptions, discrepancies, blocks, frustrations, novelties, and so forth, are occasions for ANS activity.

Discrepancies may occur for a variety of different expectations. We are rarely operating under the guidance of a single schema. Schemas are arranged hierarchically, and a number of them are active at the same time; thus, when working on a mathematical problem, one expects, for example, that the pencil one is using will not break, that one can remember the meaning of symbols, and that the functions of symbols within a particular equation will be the usual ones. In addition, certain parallel expectations (schemas) lead one to expect that the electricity will not fail, that one's chair will not collapse, that, in short, the world is more or less stable.

Whether or not an emotional construction follows the arousal produced by a discrepancy depends on the evaluative activity of the individual. It is the concatenation of an evaluative process and ANS arousal that produces emotion. Note that I am not talking about conscious appraisals of the situation when I call on evaluative processes. The appraisals may be conscious, but they typically appear in consciousness only as components of the holistic emotional experience.

We now have some rather convincing evidence that any sort of discrepancy produces autonomic arousal. One study performed by Yoshi Nakamura in our laboratory presented subjects with very simple, boring stories and asked them to imagine how the stories would end. The stories were presented one sentence at a time, and the last sentence was either consistent or discrepant with the preceding context. In all cases, we obtained slight, but significant, increases in heart rate variability following discrepancy. In a study of an interactive computer game, we got large increases in heart rate on the order of 10 to 25 beats per minute in response to unexpected events.

The effects of situational or life stress are excellent examples of unexpected events that produce visceral arousal, negative or positive evaluations, and emotional experiences. Berscheid (1983) has imaginatively described the conditions of interpersonal interactions that lead to interruptions and discrepancies and, therefore, to emotional reactions. When a relationship is meshed (when the actions of one individual depend on the actions of the other), then the two people involved may become occasions for each other's interruptions. The actions of the other are essential for one's own action; thus, emotional reactions are more likely in such meshed relationships than they are in "parallel" relationships, in which the actions of two individuals are not interdependent. A similar analysis can be applied to the relationship between task and learner. When the two are meshed, when any step by the learner implies a particular serial dependency in the task, then interruptions will have severe consequences. In a more parallel situation, however, the learner will have available diversions and alternatives that are not entirely constrained by the task.

Finally, the construction of emotions requires conscious capacity. The experience of emotion is, by definition, a conscious state and, thus, pre-empts limited capacity. Limited capacity refers to the fact that conscious contents are highly restricted at any one point in time. Whenever some particular construction pre-empts conscious capacity, then other processes that require such capacity will be impaired. The best example of this is found in stress and panic reactions, when emotional reactions prevent adequate problem-solving activities. As a result, emotional experiences are frequently not conducive to the full utilization of the cognitive apparatus; thought may become simplified (i.e., stereotyped and canalized) and tend to revert to simpler modes of problem solving. The effects are not necessarily intrusive and deleterious, however. In part, it will depend on other mental contents that are activated by the emotional experience and that may become available for dealing with situations. The relationship of emotions to discrepancies and ANS recruitment also points to their adaptive function; emotions occur at important times in the life of the organism and may serve to prepare it for more effective thought and action when focused attention is needed.

I have discussed the interactions of stress and cognitive efficiency at length in *Mind and Body* (1984). A survey of the literature supports the conclusion that stress tends to decrease attention to peripheral events and to focus attentional conscious capacity on those aspects of the situation that the individual considers important. But there lies the rub. What a particular individual considers to be important may not lead to a solution of the current problem; thus, stress can be helpful in screening out irrelevancies, but it may also result in focusing on the wrong strategy. In order to handle stress and affect effectively in a problem-solving task, the individual must be equipped with adequate knowledge of the task and possible ways of solving it. In other words, inadequate information leads to stress, but the well-informed individual can use stress constructively.

It has been argued that discrepancy is vacuous as an explanatory argument because discrepancy, as usually defined, is practically always present. To some degree, all events are somewhat discrepant from what is expected; the world changes continuously. This is correct, of course, and I would expect that some degree of arousal is present in many, possibly most, day-to-day situations, but so is some degree of feelings or moods. In fact, the criticism has been made that theories such as the cognitive/arousal theory are too discrete, that they do not account for the pervasiveness of moods and emotions, and that human beings are characterized by some feeling or mood state much of the time. The pervasiveness of discrepancies, which clearly cannot be denied, accounts for the continuous feeling states; however, the degree of discrepancy is usually slight, and the amount of arousal is small, which accounts for background feelings and moods. So-called "true" emotions typically occur with high degrees of arousal and are frequently associated with extreme discrepancies and interruptions.

I need to dwell a little on the positive emotions – the joys and pleasures of learning and problem solving, for example. Many otherwise sympathetic consumers of discrepancy theory feel uncomfortable with the notion that discrepancies are the source of autonomic arousal for positive as well as for negative events. One of the reasons for this discomfort is the common notion that discrepancy is somehow itself a negative event and is associated with frustration and other similar concepts. It is not, of course, within the confines of this theory. Consider the joys of young love. One has met the person of one's dreams and hopes, but reciprocity is not quite apparent. One is to meet again a few days hence; as the object of passion appears at the designated time, joy floods the lover, and ecstasy is near. What is discrepant? I would argue that the anticipation of the event is never devoid of doubts and fears (Will the loved one appear at the appointed time? Is he or she at all interested? Does he or she look as desirable as one has imagined?). The world of romantic love is full of such ambivalences, and whenever there are ambivalences, the actual event will be discrepant with some of them. There is no argument about the emotional quality, the "value," of the love; what is at stake is whether there is accompanying autonomic arousal generated by interruptions and discrepancies. In contrast, consider a positive event that is fully anticipated, in all its details and nuances. A prize has been won, and the check arrives in your mail. The value is still there, but the intensity will be relatively low. I believe that an appropriate analysis of positive events will disclose the operation of many ambivalent expectancies that are more than sufficient to explain the intensity of positive emotions. Of course, similar ambivalences operate for many negative events. For every expected good, there are thoughts of disappointments and slip-ups; for every expected bad, there are hopes of redemption and relief.

Similar arguments apply to the pleasures of learning, and to mathematics, in particular. For example, when a youngster first encounters mathematics, it may be strange and unusual, itself a source of arousal. But the mastery of the subject, the mastery of the novel, shows it to be not as strange and possibly frightening as it first appeared; the discrepancy is positively resolved, and positive emotions emerge. Finding solutions to problems that at first seem forbidding shows a simi-lar concatenation of events. New ideas are usually discovered in the context of uncertainties; their effect can, as we all know, be quite exhilarating—for some of us, as exhilarating as the joys of young love.

This brief review should convey the flavor of my theoretical position. My major interest here is to focus on the learning process as it generates discrepan-cies and interruptions—mainly in the production of errors and unexpected suc-cesses, as well as in the values (the evaluative reactions) that may arise in the course of the learning process.

Before I describe an application of discrepancy theory to problem solving, I wish to discuss an older tradition in the affect/learning domain—the exploration of the covariation of individual differences and task performance.

## Macroanalysis: Individual Differences and Cognitive Efficiency

One traditional approach to the relation between emotion and cognitive func-tioning has been to look at the way in which variation in some affective charac-teristic of individuals covaries with one or the other measure of their cognitive functions. I call this the *macroanalytic approach* because it typically generates global measures of both individual variation and task performance. The central interest is in the individual's performance on a task or test as a whole, which is often expressed in a single score on the task. There is little room in such an approach for a distinctly different goal—that is, how to design tasks for people so that performance can be optimized, or at least minimally affected by people's nonadaptive preoccupations and predilections (such as test anxiety). The general loss of appeal of the macroanalytic approach in recent years was, of course, influenced in large part by the rather small effects of the covariations between affect and cognitive performance (i.e., correlations of about .30, accounting for less than 10% of the variance on the problem tasks).

I first became involved with the macroanalytic approach 35 years ago, when Seymour Sarason and I studied the relation between scores on a test-anxiety scale and various memorial and problem-solving tasks (Mandler & Sarason, 1952).[3] I wrote the epilogue on my participation in that effort in 1972, pointing out the often atheoretical search for correlations that had typified the field and, more important from my point of view, the need to step back and look at (or better, *for*) theory as our primary concern. I called for more investigation into the actual operation of the processes that generate helplessness and disorganization (Mandler, 1972). In that same volume, I also suggested that test-anxiety scales may well be renamed *test-relevant-self-instruction scales* to document the fact that individuals' anxiety arises out of their specific interpretation and use of events as they unfold in the course of a test. Microanalyses of affect and cognition were needed – that is, what happens specifically in the interaction between the individual and the problem-solving task.

I want to raise one issue about the problem of individual differences in affect as it applies to complex thinking. One macroanalytic approach is to develop a test to measure individual variation and then to determine how performance on a variety of tasks covaries with that measure across a group of typical, or not-so-typical, individuals. Another approach, which is used very infrequently, is to assemble a population of representative tasks (rather than individuals) and to determine the extent to which each of the tasks produces affective reactions on a group of individuals. That produces a ranking of tasks rather than individuals, which is a quite different metric. In the best of designs, one would want to use both approaches, thus ranking both people and tasks. I believe we need to think more analytically about both kinds of approaches because a correlation of characteristics across individuals and a correlation of the *same* characteristics across tasks may produce very different results (e.g., Mandler 1959); for example, we can ascertain degree of test anxiety for a group of individuals, or we can develop an array of tasks that differ in the degree to which they elicit test anxiety. The correlation between test anxiety and effective problem-solving strategies may be negative for individuals, but nonexistent or even positive for the tasks. Which of these correlations turns out to be the one that we find important in designing curricula or teaching methods will depend on the problem that we wish to solve and the theory that informs our approach. We may be interested in designing an array of tasks; we may be interested in the melioration of individuals' difficulties with the tasks; or, optimally, we may be interested in some combination of the two. We cannot assume, however, that a particular correlation (e.g., for individuals) is an absolute measure of covariation of the two characteristics.

## Microanalysis: Discrepancy Theory and Human Error

Any discrepancy in the course of problem solving represents a potential affective episode. As I noted earlier, such episodes must be seen within the context of a general flow of affective and moodlike changes. However, it is possible to identify specific kinds of discrepancies and interruptions that may occur in the course of problem solving, in general, and in mathematical reasoning and learning, in particular. The major class of such events can generally be termed *errors*, when the learner does something or thinks something that is different from his or her original intention or is different from what ought to happen. I shall discuss errors as such shortly, but we must keep in mind that people frequently engage in actions that they believe to be correct (i.e., they proceed as intended, but the actions are in fact false, incorrect, illogical, etc.). In the case of unintended errors, the discrepancy arises because of a mismatch between what is intended and what occurs; in the other cases, the mismatch is between an expected outcome ("I thought I did what would solve the problem") and the real-world response ("It didn't work"). Most intended and unintended errors are coupled with a negative evaluation of the current situation. These errors will usually result in negative affect (unhappiness, disgust, despair) in varying degrees of intensity. The result is interference with ongoing cognitive processes because of both the pre-emption of conscious capacity and the search for correction (looking for the source of the disruption). If this process is allowed to go unchecked (i.e., if the learner is continuously producing errors), the intensity of negative affect will increase. Such a sequence may eventually produce: abandonment of the task because it is too noxious to tolerate, panicky quasi-random attempts at solution, and general disorganization.

The other general class of discrepancies may be termed *successes*, which also may be intended or unintended. The intended success is a step in the solution process that works; the unintended success occurs when a particular action or thought produces unexpected positive results ("I just tried that because I couldn't think of anything else, and it worked"). In general, it is likely that the smooth progression of some planned course of problem solving will produce little in the way of arousal. After all, if I go through some cookbook steps toward solution, the expectation is that each step will be successful, and no discrepancy occurs; however, when I am not sure of the successive steps, success may be slightly arousing. During the early stages of the learning process, when the learner is unsure, we expect—and, of course, find—more affective states and more inter-ference with learning. The well-practiced solution runs off smoothly and, by definition, produces little affect. From a more global perspective, the final solu-tion of a problem will produce discrepancy if the learner is unsure of his or her ability; joy, delight, and satisfaction result.

Norman (1981) analyzed action slips within the general context of schema the-ory. Well-learned action sequences (and by implication, thought sequences) are specified by a high-level schema (the intention), which then recruits lower level schemas that guide the various components of the developing sequence. These component schemas are presumably activated both by the original intention (and its descendants) and by appropriate conditions of the task as the sequence (the problem solving) develops—that is, the thought sequence emerges interactively from both top-down and bottom-up sources. Norman discusses three categories of slips: errors in the formation of an intention, defective activation of compo-nent schemas, and defective triggering. In another paper, Norman (1980) makes the distinction between mistakes, which result from errors in the formation of an intention, and slips, which are errors in the execution of an intention. I cannot discuss these papers in detail, but they are worthy of attention for the purpose of analyzing specific tasks and the kinds of mistakes and errors that can occur.[4]

## Errors: What Are They Good For?

The typical work on human errors and slips pays hardly any attention to problems of the consequences of the errors and essentially no attention to any affective consequences. Norman (1980), for example, is much concerned with problems of limited conscious capacity (he refers to it as *short-term-memory* capacity), but the possibility that affective consequences of errors demand and pre-empt such capacity is not seriously entertained. Because I am not concerned here with the causes or categorization of errors, but rather with their affective consequences, I want to emphasize exactly that point. The affect that occurs when errors or mistakes are made is a conscious event; it pre-empts our limited conscious capacity and has several consequences. First of all, just the perception of autonomic arousal, when severe, is demanding enough to hinder one's ability to act effectively ("I am so upset, I can't think straight"). Second, the occurrence of strong negative affect produces an immediate attempt to remove the reasons (the cause). That may be a useful consequence when the removal of the offending event is more important than a continuation of the problem-solving activity. In the usual learning context, however, the latter may be more adaptive. Third, even a positive affective experience in the context of problem solving may be deleterious. The individual solves some subroutine and stops, in part to admire the achievement, in part to savor the affect. In the process, the current contents of consciousness are displaced; one loses one's place in the stream of problem solving.

Is it the case, then, that errors and mistakes should be avoided whenever possible, and that tasks should be designed so that problem solving can be learned without errors? I now come to the subject of errorless learning: Can it happen? Is it a good thing? I shall return to the real world soon and discuss the inevitability of errors and mistakes and the advantages of having had prior experiences with them.

The concern with errorless learning started about 25 years ago when Skinner (1961) argued that learning should involve as few errors as possible, and that it should be possible (e.g., by making incremental steps small enough) to acquire skills and knowledge without the nuisance and debilitating effects of errors. Skinner provided little, if any, empirical evidence for his argument, but Terrace (1963, 1972) demonstrated an analogue of human errorless learning in animal discrimination training. The animal is required to distinguish between a positive stimulus and a negative stimulus. The required response must be made when a positive stimulus is presented and omitted when a negative stimulus is presented. Terrace showed that by slowly bringing in the negative stimulus, the animal (typically the pigeon) could learn to discriminate between a positive stimulus and the negative stimulus without making many (or any) responses to the negative stimulus. The major claim about errorless learning that concerns me here is that one of the by-products of discrimination learning does not appear following errorless learning. That by-product is emotional (and aggressive) behavior when a negative stimulus is presented.[5] It has been known for quite some time that negative stimuli (and the absence of reinforcement during extinction) produce emotional behavior in animals. A major theoretical position in another branch of behaviorist theory—Amsel's frustration theory—addresses these effects (e.g., Amsel, 1962).[6] If a method could be found that eliminated the emotional consequences of negative stimuli (and their avoidance), it would have important consequences for theories of learning and teaching; however, that is not the case. Rilling and his students (see Rilling, 1977, for a review) showed that most of the by-products of discrimination learning, and in particular aggressive emotional behavior, also occur following errorless learning. These behaviors "bear little relationship to the occurrence or nonoccurrence of errors during S- [negative stimulus]" (Rilling, 1977, p. 467). In keeping with the expectancy notions of interruption theory, "withdrawal of opportunity for reinforcement is one of the primary determinants of extinction-induced aggression" (Rilling, 1977, p. 469). It appears that it is the absence of positive stimuli (the expectations developed during the errorless responses to the positive stimulus) that produces arousal and emotional behavior.[7]

I am not advocating that one look to the animal learning literature for an understanding of human cognitive processing; however, the only data available suggest that errorless learning does not produce an affect-free learning environment. If we assume that there are, at present, no methods that produce painless learning and that the consequences of learning, no matter how painless, may still produce affective reactions, how do we accommodate such a state of affairs?

My main argument should be obvious by now. I do not believe that it is possible for an individual to live a cocooned life in which no failures, no mistakes, no slips will be encountered. These missteps in life (including mathematical learning) can be highly deleterious if totally unexpected; they are less interfering and less intense if their occurrence is a normal feature of life and, therefore, expected to occur at some time or another. The occurrence of affective reactions to learning experiences, arousal due to discrepancies during the learning process, and other bad experiences build up expectancies, and are built up into experiential schemas. Events that are totally unexpected produce maximal arousal and affective reactions. If, however, we know that errors (and very specific errors) can occur in problem-solving situations, the errors are not surprising and their consequences are not unexpected. The cocooned individual has extreme reactions to novel situations. In short, I believe that it is advantageous—at least in the problem-solving situation—for an individual to have been exposed to the "school of hard knocks." There is a nice set of demonstrations available in the animal (and human) experimental literature on the partial reinforcement effect. If a response is consistently reinforced during learning, then extinction of that response once reinforcement is absent will be relatively rapid; however, if the initial reinforcement is only given on some percentage of the acquisition trials, then extinction of the response will be slow. It is the experience with unreinforced responses that makes them less effective during extinction.[8]

How, then, do we design learning and teaching tasks? It is here that the notion of microanalysis receives its most direct application. We need to analyze the task and the learner to ensure that the surprises, errors, and missteps during acquisition are relatively minor — that is, that they can be mastered by some alternative route, by substitute actions, or by a restatement of the problem (one's intentions). At the same time, we do not want to avoid such affective incidents altogether; we want the learner to be prepared for them in the future. Thus, small incremental steps, specific instruction in subroutines, attention to possible difficulties, and instructions for the anticipation of difficulties play a role in analyzing and constructing a task. We also need to take into account the individual's attitudes and beliefs about the problem, because they will interact with the expectations that will be developed and perhaps be confirmed or violated.

## Do We Want Affectless Learning?

The major message of the foregoing section is that affectless learning is not a possible goal for a theory or for the praxis of instruction. Common sense tells us that emotions and affective reactions are with us now and forever. Not surprisingly, the usual measures of individual differences tell us the same thing. The macroanalytic methods, no matter how unsatisfactory, do give us some indication of the person's affective preoccupations with tests, cognitive tasks, and similar situations. These dispositions will interact with the emotional consequences of specific errors, failures, successes, and strategies; thus, one would expect an individual with a high level of anxiety or arousal to produce a more debilitating emotional reaction than a more placid, unresponsive person. In the analysis of a task and in the design of learning, one should pay attention to such factors and design different (less demanding) acquisition paths for the more emotionally reactive learners.

Apart from the inescapable fact of life that emotions and affect are an integral part of what we call human nature, what we recognize as the constitutive aspects of our humanity, one can argue that an affectless learner would not learn much. Such a conclusion arises out of a brief consideration of the relation between emotion and motivation. Research in so-called "motivational" variables has fallen upon hard times in recent years. One reason for this is that the new cognitive orthodoxy has not yet found a reasonable mechanism for incorporating motivation into its theoretical superstructure.[9] One possible direction for this research would involve more extensive use of emotional states for motivational arguments. Suggestions proposed over the years have been essentially hedonistic ones — for example, people prefer to generate positive subjective (emotional) states and avoid negative ones. A corollary is that conditions (situations, thoughts, actions) that give rise to positive states will be sought out or repeated, and those that generate negative states will be avoided. Such a neobehaviorist position is probably the best we can do at present. At the descriptive level, it accounts for such phenomena as students' aversion to mathematics as a result of early unhappy experiences. It also brings the emotional domain into the important area of students' beliefs about strategies and tactics. Schoenfeld (1985; see also Silver, in the same volume) has argued persuasively that beliefs and metacognitions are important contributors to problem-solving performance. The extent to which such beliefs are confirmed or disconfirmed will, in turn, generate positive or negative emotional states, which will then feed back to the degree to which these beliefs may be maintained or discarded.

It is beyond the scope of this chapter to discuss the social and cultural factors that determine one's attitudes and beliefs about problem solving. I do want to recognize the importance of such factors, because they determine many of our expectations about: situations and tasks, about the social interactions that are important in the learning and teaching situations, and about our successes and failures and their relevance to our social functions and our self-image.

The argument has been advanced (e.g., Toda, 1982) that our emotions evolved to fit an environment that was wild, dangerous, and uncivilized. It is argued that we now live in a different environment, but with emotions that are outmoded and fit different situations. Wald (1978), for example, has averred that large portions of our animal heritage "have become inappropriate to civilized life" (p. 277). Wald focuses on the "violent" emotions, which he lists as pain, fear, and rage, which he considers "out of place," but he forgets other violent emotions, such as the positive passions of lust, love, ecstasy, and joy. Are these, too, out of place?

What is often forgotten is that the environment to which we originally "adapted" by developing these emotions was designed by nature, chance, or whatever; however, the very environment for which we are said to be unfit emotionally was designed by, guess who, the organism that does not fit it. I wonder whether an argument that says that it is unlikely that cultural evolution would design an environment precisely unfit for the designer is not just as viable as the argument that says we are, in fact, unfit for the present environment. Besides, in what way are we unfit?

Both affect and learning are characteristics of contemporary mankind. As psychologists, we wish to understand them, and, whenever possible, modify and make adjustments to them in order to improve the human condition. We want learning to be painless and fast, and affect to be joyful and occur when wanted. We cannot have either of these states. The best we can do at present is to understand how learning and affect come about, how they interact, and how their inevitable symbiosis can be put to the use of our students and our society.

*Acknowledgment.* Preparation of this chapter was supported in part by grants from the Spencer Foundation and the National Science Foundation.

## Footnotes

[1]The discussion will necessarily be a very sketchy and inadequate presentation of the theory. It is intended to introduce some of the notions I need to use in the following sections.

[2]For an extensive discussion of classes of values, including the problem of completion, see *Mind and Body* (1984).

[3]The manifest anxiety scale, developed by Janet Taylor Spence (Taylor, 1953), made its appearance at about the same time and had a much greater effect on ongoing covariational research.

[4]Other important contributions to error theory are Reason's papers (e.g., 1977, 1979).

[5]There are other claims about the by-products of errorless discrimination learning that need not concern us here.

[6]I have discussed these developments and their obvious relation to interruption theory in *Mind and Emotion* (1975).

[7]In my only excursion into animal experimentation, I demonstrated that the absence of reinforcement is a very powerful inducer of extreme emotional reactions even when the animals are satiated—that is, the very lack of opportunity for engaging in well-learned behavior has marked emotional consequences. When substitute behaviors are made available to the animals, however, the emotional behavior can be suppressed (Mandler & Watson, 1966).

[8]See my book *Mind and Emotion* (1975) for a more detailed discussion of the learning and affective consequences of partial reinforcement.

[9]See Gallistel (1980) for one structural approach to motivation.

## References

Amsel, A. (1962). Frustrative nonreward in partial reinforcement and discrimination learning. *Psychological Review, 69*, 306–328.

Averill, J.R. (1980). A constructivist view of emotion. In R. Plutchik & H. Kellerman (Eds.), *Theories of emotion* (pp. 305–339). New York: Academic Press.

Berscheid, E. (1983). Emotion. In H.H. Kelley, E. Berscheid, A. Christensen, J.H. Harvey, T.L. Hudson, G. Levinger, E. McClintock, L.A. Peplau, & D.R. Peterson, *Close relationships* (pp. 110–168). San Francisco: W.H. Freeman.

Bower, G.H. (1981). Mood and memory. *American Psychologist, 36*, 129–148.

Bower, G.H., & Mayer, J.D. (1985). *In search of mood-dependent retrieval.* Unpublished manuscript.

Broadbent, D.E. (1971). *Decision and stress.* New York: Academic Press.

Easterbrook, J.A. (1959). The effect of emotion on cue utilization and the organization of behavior. *Psychological Review, 66*, 183–201.

Gallistel, C.R. (1980). *The organization of action: A new synthesis.* Hillsdale, NJ: Lawrence Erlbaum Associates.

Isen, A.M., Means, B., Patrick, R., & Nowicki, G. (1982). Some factors influencing decision-making strategy and risk taking. In M.S. Clark & S.T. Fiske (Eds.), *Affect and cognition: The 17th Annual Carnegie Symposium on Cognition* (pp. 243–261). Hillsdale, NJ: Lawrence Erlbaum Associates.

Kahneman, D. (1973). *Attention and effort.* Englewood Cliffs, NJ: Prentice-Hall.

Lazarus, R.S., Kanner, A.D., & Folkman, S. (1980). Emotions: A cognitive-phenomenological analysis. In R. Plutchik & H. Kellerman (Eds.), *Theories of emotion* (pp. 189–217). New York: Academic Press.

Mandler, G. (1959). Stimulus variables and subject variables: A caution. *Psychological Review, 66*, 145–149.

Mandler, G. (1972). Helplessness: Theory and research in anxiety. In C.D. Spielberger

(Ed.), *Anxiety: Current trends in theory and research* (Vol. 2, pp. 359–378). New York: Academic Press.

Mandler, G. (1975). *Mind and emotion*. New York: Wiley.

Mandler, G. (1984). *Mind and body: Psychology of emotion and stress*. New York: Norton.

Mandler, G., & Nakamura, Y. (1987). Aspects of consciousness. *Personality and Social Psychology Bulletin, 13*, 299–313.

Mandler, G., & Sarason, S.B. (1952). A study of anxiety and learning. *Journal of Abnormal and Social Psychology, 47*, 166–173.

Mandler, G., & Shebo, B.J. (1983). Knowing and liking. *Motivation and Emotion, 7*, 125–144.

Mandler, G., & Watson, D.L. (1966). Anxiety and the interruption of behavior. In C.D. Spielberger (Ed.), *Anxiety and behavior* (pp. 263–288). New York: Academic Press.

Norman, D.A. (1980). *Errors in human performance* (Report No. 8004). San Diego: University of California, Center for Human Information Processing.

Norman, D.A. (1981). Categorization of action slips. *Psychological Review, 88*, 1–15.

Reason, J.T. (1977). Skill and error in everyday life. In M. Howe (Ed.), *Adult learning: Psychological research and applications* (pp. 21–44). London: Wiley.

Reason, J.T. (1979). Actions not as planned: The price of automatization. In G. Underwood & R. Stevens (Eds.), *Aspects of consciousness: Vol. 1, Psychological issues* (pp. 67–89). London: Academic Press.

Rilling, M. (1977). Stimulus control and inhibitory processes. In W.K. Honig & J.E.R. Staddon (Eds.), *Handbook of operant behavior* (pp. 432–480). Englewood Cliffs, NJ: Prentice-Hall.

Schoenfeld, A.H. (1985). Metacognitive and epistemological issues in mathematical understanding. In E.A. Silver (Ed.), *Teaching and learning mathematical problem solving: Multiple research perspectives* (pp. 361–379). Hillsdale, NJ: Lawrence Erlbaum Associates.

Silver, E.A. (1985). Research on teaching mathematical problem solving: Some underrepresented themes and needed directions. In E.A. Silver (Ed.), *Teaching and learning mathematical problem solving: Multiple research perspectives* (pp. 247–266). Hillsdale, NJ: Lawrence Erlbaum Associates.

Skinner, B.F. (1961). Why we need teaching machines. *Harvard Educational Review, 31*, 377–398.

Taylor, J.A. (1953). A personality scale of manifest anxiety. *Journal of Abnormal and Social Psychology, 48*, 285–290.

Terrace, H.S. (1963). Discrimination learning with and without errors. *Journal of the Experimental Analysis of Behavior, 6*, 1–27.

Terrace, H.S. (1972). By-products of discrimination learning. In G.H. Bower (Ed.), *The psychology of learning and motivation* (Vol. 5, pp. 195–265). New York: Academic Press.

Toda, M. (1982). *Man, robot, and society*. Boston: Martinus Nijhoff.

Wald, G. (1978). The human condition. In M.S. Gregory, A. Silver, & D. Sutch (Eds.), *Sociobiology and human nature* (pp. 277–282). San Francisco: Jossey-Bass.

Yerkes, R.M., & Dodson, J.D. (1908). The relation of strength of stimulus to rapidity of habit-formation. *Journal of Comparative and Neurological Psychology, 18*, 459–482.

Zajonc, R.B. (1980). Feeling and thinking: Preferences need no inferences. *American Psychologist, 35*, 151–175.

# 2
# The Role of Affect in Mathematical Problem Solving

Douglas B. McLeod

When students are given a nonroutine mathematical problem to solve, their reactions often include a lot of emotion. If they work on the problem over an extended period of time, the emotional responses frequently become quite intense. Many students will begin to work on a problem with some enthusiasm, treating it like a puzzle or game. After some time, the reactions become more negative. Students who have a plan to solve the problem may get stuck trying to carry out the plan. They often become quite tense; they may try to implement the same plan repeatedly, getting more frustrated with each unsuccessful attempt. If the students obtain a solution to the problem, they express feelings of satisfaction, even joy. If they do not reach a solution, they may angrily insist on help so that they can reduce their frustration.

Given the intensity of the emotional responses to problem-solving assignments, it is a bit surprising that research on mathematical problem solving has not looked into affective issues more seriously. Since the appearance of *An Agenda for Action* (National Council of Teachers of Mathematics, 1980) and its recommendation for greater emphasis on problem solving, a substantial amount of effort has gone into research and related development of materials to teach problem solving at all ages and all ability levels (Shufelt, 1983). However, neither research nor development has provided teachers with much guidance on how to deal with affective issues.

Recent research has made substantial progress in characterizing the cognitive processes that are important to success in mathematical problem solving; however, the influence of affective factors on these cognitive processes has yet to be studied in detail. The purpose of the chapter is to propose a theoretical framework for investigating the affective factors that help or hinder performance in mathematical problem solving.

## Related Research on Mathematical Problem Solving

Although most research on mathematical problem solving has proceeded out of the Polya (1945) tradition, giving little attention to affective issues, reviews of research on teaching problem solving have noted the importance of affective factors (Lester, 1983). When researchers write for teachers, affective concerns are addressed (e.g., Kantowski, 1980), but even researchers who recognize the importance of affect do not generally include affective factors in their studies. This reluctance to include affective factors is due in part to the lack of a suitable theoretical perspective.

When studies have included affective factors, the data appear to be quite promising. Charles and Lester (1984), for example, reported that instruction in problem solving produced positive changes in confidence on the part of both students and teachers in Grades 5 and 7. Even though affective factors were not the major focus of the study, it was possible to see changes in affective factors and in the influence of these factors on problem-solving performance.

Other researchers have also noted the importance of affective factors in problem solving. Silver (1982), in his discussion of metacognitive aspects of problem solving, discussed the relationship of affective factors (like confidence and willingness to persist) to problem-solving performance. His speculations suggest that these affective factors may have a substantial effect on the metacognitive processes of problem solvers.

Schoenfeld (1983), in a similar way, has written about the management and control systems that problem solvers use as they address mathematical problems. In his research, he emphasizes the role that belief systems play in determining the kinds of managerial decisions that problem solvers make. He suggests that attitudes toward mathematics and confidence about mathematics may be aspects of student belief systems that have an important effect on how students manage their cognitive resources.

In a recent conference devoted to research on teaching mathematical problem solving, Silver (1985) noted the need for more research on affect. Earlier efforts have suffered from the lack of a theoretical perspective on affective issues, and from the choice of general measures like attitude and interest in problem solving, rather than more specific factors (frustration, satisfaction) and their relationship to the cognitive processes used in problem solving. Silver suggests that the time is right to renew our efforts to identify the relationships between affect and problem solving.

## Curriculum Development and Affect

Materials designed to teach problem solving may mention the importance of attitude and motivation, but they do not address affective issues in a systematic way. Some problem-solving books for secondary school students (e.g., Adams, 1980; Whimbey & Lochhead, 1982) include discussions on confidence and persistence, but place little emphasis on such topics. Materials for elementary school students (e.g., Ohio Department of Education, 1980) sometimes address the same kinds of general issues, but usually in the comments to teachers. In many cases, affective issues are just ignored.

Some books do provide thoughtful and direct discussions of affective issues in problem solving. Mason, Burton, and Stacey (1982), for example, address in some detail the difficulties involved in getting "stuck" when trying to solve a problem. The authors give students ways to deal with their feelings of frustration when they are stuck and suggest useful heuristics that will help get the students moving again. They note that expressing frustrations can be helpful, and they discuss strategies for building confidence. They emphasize positive emotions; for example, they talk about the pleasure that comes from making conjectures, as well as the joys of finding solutions to challenging problems. Mason, Burton, and Stacey (1982) do not develop a psychological theory to support their recommendations regarding affective issues. Burton (1984), however, provides some information on the philosophy and experience that influenced their work.

In summary, both researchers and developers agree that affective issues are important in mathematical problem solving; however, little has been done to incorporate affective concerns in a systematic way in either research or curriculum development. Because much research on problem solving is now done within the framework of cognitive science, it is appropriate to consider how affective issues could be addressed within that framework.

## Mandler's View of Mind and Emotion

As Norman (1981) points out, most cognitive theorists prefer to ignore the affective domain and concentrate instead on developing information-processing models of purely cognitive systems. There is general agreement among researchers that affect is much more complex and difficult to describe than cognition (Simon, 1982). Norman suggests that even the progress made in areas like research on memory has been marred by the sterility of a purely cognitive approach. This focus on pure cognition reduces the usefulness of these theories for mathematics education.

By concentrating on just the cognitive aspects of problem solving, researchers are employing the usual heuristic of trying to solve a simpler problem. The importance of affective issues, however, is becoming evident to increasing numbers of cognitive psychologists. In cognitive science, for example, Norman (1981) lists emotion as one of the 12 major issues that need to be addressed in future research. A major influence on Norman's thinking is the work of George Mandler.

Mandler, a cognitive theorist, has played a leading role in the development of cognitive science (Mandler, 1985). Mandler's book *Mind and Emotion* (Mandler, 1975) has had a significant impact on the field of cognitive science as well as on psychology, in general. Strongman (1978) reviewed current theories related to the psychology of emotion and indicated that Mandler's views were among the most influential.

Mandler's recent work (1984) refines his earlier book and extends his ideas in ways that make his perspective more applicable to mathematical problem solving. His approach to research on emotion is from the perspective of the "new" cognitive psychology, the information-processing approach that is common to most research on problem solving. From his point of view, a major source of emotion is the interruption of an individual's plans or planned behavior. In psychological terms, these plans result from the activation of a schema that produces an action sequence, and this sequence has "a tendency to completion" (Mandler, 1984, p. 173). When an interruption occurs, the normal pattern of completion for these sequences of thought or action cannot take place. The result of the interruption is the physiological arousal of the individual. This arousal might be observed as muscle tension or rapid heartbeat. Along with the arousal, the individual evaluates the meaning of the interruption. The result of the evaluation is then interpreted as surprise, frustration, joy, or some other emotion.

These interruptions of planned, organized sequences of thought or action are also referred to as blockages, or discrepancies between what was expected and what was perceived. Although there is more to Mandler's general framework than just the notion of interruption, this idea is central to his development of a psychology of emotion. His theory is a general one, and it has been applied in a variety of situations. It seems to be particularly appropriate to the analysis of affective issues in problem solving.

Researchers on problem solving usually define a problem as a task in which the solution or goal is not immediately attainable; there is no obvious algorithm for the student to use. In other words, the student's initial reaction to the problem is that no solution is evident; the problem solver is blocked. The initial plan to solve a nonroutine problem is often inadequate; the plans are interrupted. New strategies, or heuristics, may be applied. In other words, this definition of a problem is exactly the situation that Mandler uses to describe how interruption and arousal lead to emotion.

In Mandler's view, the cognitive evaluation that is combined with the arousal can result in either a positive or a negative emotion. We see both in the reactions of students to problem solving. What is exciting for one student can be depressing to another. Either kind of emotion can result from the same type of interruption; therefore, the way that students interpret the effect of the interruption is very important.

Mandler points out that the intensity of the reaction to the interruption is related to the degree of organization of the student's mental activity. In mathematics, in which students spend so much of their time doing routine exercises, student actions are very highly organized; students expect to complete most mathematical tasks without any blockages or delays. Thus, the blocks that inevitably interrupt problem-solving activities may lead to intense emotions. The fact that this intensity can actually be observed in problem solvers is another piece of evidence that Mandler's views on emotion may be particularly useful for research on mathematical problem solving.

This brief introduction to Mandler's ideas on emotion cannot do justice to his theoretical position (Mandler, 1984); however, it should make clear the close connection between his ideas on emotion and current research on problem solving.

# Some Aspects of Cognition and Their Relation to Affect

Problem solving is clearly a cognitive activity. Because there is no established catalog of the cognitive processes that are most important in problem solving, this section summarizes several major categories that are relevant to affective issues. It is the thesis of this chapter that the cognitive processes involved in problem solving are particularly susceptible to the influence of the affective domain.

## *Memory and Representation Processes*

Although some progress has been made in research on memory, Mandler (1984) notes that we have little data regarding affective influences on storage and retrieval processes. The importance of these processes to problem solving is clear; retrieval from long-term memory, for example, is often crucial, and the failure of retrieval processes under stress is a major difficulty in problem solving.

Bower (1981) reports that mood (affect) can influence memory processes — to oversimplify, a positive mood helps with recall of pleasant experiences; however, this research is typically based on moods that are considerably less intense (and involve less arousal) than the emotions involved in mathematical problem solving.

Along with memory, representation processes play a crucial role in problem solving. Mathematicians certainly display preferences in representational styles and sometimes classify themselves as either geometric or algebraic in the way that they address problems. Students also develop preferences for certain problem-solving strategies; some seem reluctant to draw figures, for example. Affective influences on these representational styles remain unexplored.

## *The Role of Consciousness*

The resurgence of interest in consciousness seems to be a result of the movement from behaviorism to information-processing theory among psychologists. Norman (1981) lists the study of consciousness as 1 of 12 important topics for cognitive science research. Mandler (1975, 1984) argues that the concept of consciousness is needed to account for important aspects of human information processing; for example, notions of limited processing capacity and retrieval from long-term memory make little sense without a corresponding notion of consciousness.

Rigney (1980) discusses aspects of consciousness that seem especially relevant to affective issues in problem solving. He notes some of the differences between conscious and unconscious mental processing. He suggests, for example, that conscious processing is rather slow when compared with unconscious processing, and he contrasts the limited capacity of consciousness and working memory with the unconscious, which does the major part of all mental processing. Rigney's ideas on unconscious processes are especially interesting for mathematical problem solving because they are so much in line with Hadamard's (1945) ideas on mathematical discovery.

Even though the strict behaviorism of the past is no longer with us, there is still a substantial amount of distrust among psychologists regarding research on conscious and unconscious processes. Mandler (1984) notes that researchers sometimes seem to use short-term memory as a more acceptable substitute for consciousness. Wickelgren (1974) dismisses the possibility that the unconscious might play a significant role, at least in retrieval processes. Perhaps recent work in cognitive science by Johnson-Laird (1983) will help to legitimize research on this topic. From Johnson-Laird's point of view, the mind's "operating system" involves a network of parallel processors, and this parallelism leads in a fundamental way to the differences between conscious and unconscious processing. In any case, we are clearly at an early stage in our understanding of consciousness, and further investigation of this important concept is needed, especially in relation to the cognitive processes involved in problem solving.

## The Role of Metacognition

Kilpatrick (1985) has also explored the concept of consciousness in his stimulating paper presented at the Fifth International Congress on Mathematical Education. He suggests that we need to help students analyze how consciousness operates and how they can manage their own mental resources. He refers to this as *metacognition*. Certainly, metacognition is important to success in problem solving, in which the solver must make decisions about which strategy to apply and how long to keep on trying it before stopping and selecting a new strategy (Silver, 1982; Schoenfeld, 1983). If we can help problem solvers reflect on their own cognitive processes, we should also be able to bring to consciousness an awareness of their emotional reactions to problem solving. An increased awareness of these emotional influences should give students greater control over their cognitive processes.

Discussion of metacognition is made more difficult because of the lack of clarity in the concept. Brown, Bransford, Ferrara, and Campione (1983) suggest that we separate the notion of metacognition into two parts, a recommendation that Lawson (1984) supports. The first part is the awareness of one's own cognitive processes, or knowledge of cognition. The second part refers to the need to make decisions about our cognitions—the need for executive processes, or the regulation of cognition. Both of these aspects of metacognition appear to be crucial in problem solving and are closely tied to affective concerns; for example, knowledge about cognition is related to expectations regarding success and failure, an important part of student confidence, and the regulation of cognition is related to issues of persistence in problem solving. Developing a theoretical perspective that incorporates these ideas is an important task for research on problem solving.

## *The Role of Automaticity*

Given the limitations of our working memory (or the limits on consciousness), most of our mental processing is done automatically, with no involvement of our consciousness or metacognition. We could not live without automaticity; for one thing, we would not be able to get in our cars and drive to work safely without being able to handle a car in a relatively automatic way. On the other hand, there are many occasions when the tendency toward automatic processing works against the needs of problem solvers. So much of the current mathematics curriculum is designed to develop automatic responses (basic facts and algorithms in arithmetic, for example) that it is not surprising to find students attacking problems in an automatic way.

Automaticity is normally a part of cognitive theories (Resnick & Ford, 1981; Shiffrin & Dumais, 1981), but the processing of affective information related to mathematics is also done automatically, in large part. Students who get frustrated by a nonroutine problem, for example, will often just quit; they assume, without really thinking about it, that frustration is a signal to go and get help, rather than a normal part of problem solving. No doubt the student has practiced this response to frustration many times, and it just happens automatically; no conscious decision is needed.

Automaticity, then, has its disadvantages as well as its advantages in problem solving. The trick is to bring automatic processes back under conscious control when you need to apply them in nonroutine situations. Interruption plays a significant role here, because the interruption of an automatic process will help bring it to conscious awareness (Mandler, 1984). Problem solvers need to be capable of bringing both their cognitive processing and affective processing under conscious control, thus allowing their metacognitive processes to direct their efforts to find a solution.

# Problem-Solving Strategies and Their Relation to Affect

Instruction in mathematical problem solving is typically based on the heuristics suggested by Polya (1945). For example, Charles and Lester (1984) provided instruction in strategies such as using a diagram, making a table, and working backwards. The relationship of these (and other) strategies to affective considerations is as yet unstudied.

It seems reasonable to assume that affective influences will vary depending on the kinds of strategies being used. Consider, for example, the problem of finding the number of diagonals in a polygon with 15 sides. Once students learn how to reduce this problem to a sequence of simpler tasks, they can see how the numerical patterns develop. Seeing the pattern not only gives students the answer, but it gives them confidence as well. If a problem can be broken down into a series of simpler tasks, students can have some confidence that they will be able to find a solution. Other problems may not be as useful for developing confidence. Problems that rely on insight, like the nine-dot, four-line problem (e.g., Wickelgren, 1974), seem to undermine confidence, as do many other "puzzle" problems. Also, geometry problems that require the use of ruler and compass may cause special difficulties for students who are uncomfortable with those kinds of tools (McLeod, 1985).

Affective influences will vary not only by strategy, but also by stages. Polya (1945), for example, lists four stages: (a) understand the problem, (b) plan your solution, (c) carry out your plan, and (d) examine your solution. The frustrations of trying to carry out an inadequate plan are likely to be quite different from the affective responses to examining your proposed solution, as discussed below.

Burton (1984) describes how affective responses to problems accompany her three phases of entry, attack, and review—phases that recur in her helix-shaped model of the problem-solving process. As problem solvers engage a problem, their curiosity is aroused. This entry phase can be followed by attack (from those who have sufficient confidence) or withdrawal (from those who do not). Finally, Burton suggests that problem solvers who find a solution have a sense of achievement that fuels their "looking back" and taking stock in a "review" phase.

Although Burton's ideas are appealing, they are not supported by theory and by empirical test. Data from Kantowski (1977), for example, suggest that students seldom look back, or review the problem solution in the way that Burton hypothesizes and that Polya recommends. Instead, once a solution has been proposed, students tend to lose all interest in the problem and go on to other tasks. Bloom and Broder (1950) provide another perspective on this point. They note how problem solving involves cycles of tension and relaxation; once a solution is achieved, the problem solver relaxes. The relaxation may occur even if the solution is incorrect. Because the blockage or interruption (in Mandler's sense) has now been removed, the problem solver reports feelings of satisfaction and feels no need to go back and analyze the problem further.

In summary, affective influences on problem solving will vary according to the kind of heuristic strategy that the problem requires and according to the phases through which the problem solver moves in addressing the problem. The roles that the affective domains play in the various aspects of problem solving are not well understood, and seem worthy of more thorough investigation than they have received up to this point.

## Dimensions of a Theory

Because problem-solving performance is heavily influenced by affective factors, research on problem solving needs to take affective issues into account. We need to postulate a theoretical framework for affective influences on problem solving that is compatible with cognitive science, because a cognitive science perspective is becoming the most common setting for research on problem solving. Mandler (1984) presents such a framework.

The theory of affective influences on problem solving will need to pay special attention to the various kinds of cognitive processes, as well as the role of consciousness and automaticity in problem solving. Also, the theory will need to address specific aspects of mathematical problem solving, including the different phases of problem solving and the different heuristics that are taught. Although the theory presented in this section will be based on Mandler's work, it will also take into account contributions from other areas, including other psychological research on affect, the influence of the social environment, the role of belief systems, and related topics.

In this section, a sketch of a theoretical framework for research on affective influences in mathematical problem solving is presented. First, some characteristics of affective states are considered. These characteristics seem to be the ones that are most relevant to instruction in mathematical problem solving. Second, the relationship of these characteristics to different types of cognitive processes, to different instructional environments, and to student belief systems is described. We conclude by discussing briefly some of the implications that might be drawn from research related to affective influences on mathematical problem solving.

## Characteristics of Affective States

The affective states of problem solvers can be described in many ways. In this section, we focus on the magnitude and direction of the emotion, its duration, and the solver's level of awareness and control of the emotion.

### Magnitude and Direction

Affective influences on problem solving vary in their intensity (or magnitude), as well as in their direction (positive or negative). The most common reaction expressed by students is the frustration of getting stuck—a reaction that is frequently intense and negative; however, students also mention positive emotions, especially the satisfaction of the "Aha!" experience, which is also perceived rather intensely. Other reactions to problems, such as liking them because they have an applied, real-world flavor, seem to be less intense than frustration or satisfaction.

Observations in junior high school classrooms indicate the importance of positive emotions to both students and teachers. A teacher in a seventh-grade arithmetic class reported that she spent the first several weeks of the new school year assigning very easy, but nonroutine, problems to her students. After these successful experiences in problem solving, her students would persist in solving problems for a substantial period of time. Another junior high teacher reported his dissatisfaction with problem-of-the-week assignments in a pre-algebra class; his dissatisfaction stemmed from the fact that his students would not attempt problems over a period of time as long as 1 week. The first time he assigned such a problem, all but two students quit after the first day. By changing to simpler problems (to be solved in 1 day), he found that he could give more students the experience of success and help them enjoy the satisfactions of problem solving.

Specifying these basic characteristics of affective influences on problem solving is important. It helps remind us that we need to emphasize the positive emotions that are a part of problem solving. Individuals who enjoy the challenge of mathematical problem solving are not really typical; their reaction to challenging problems is very different from that of students in a normal classroom. Students need help in experiencing the joys of setting up conjectures (Mason, Burton, & Stacey, 1982), the art of problem posing (Brown & Walter, 1983), and the social rewards that come from group work on solving problems. Instruction in problem solving should help students deal with negative emotions, such as the inevitable frustrations of trying to solve nonroutine problems. Instruction in problem solving should do more than that, however. It should also provide those positive experiences that help students enjoy problem solving.

## DURATION OF THE EMOTION

Another important characteristic of affective states is the duration of the emotion (Kagan, 1978). Affective reactions during problem solving are typically intense, but they are relatively short in duration. Students have difficulty persisting in problem solving if their reaction is intense and negative, so they tend to quit and reduce the magnitude of the emotion. Students who persist will go from positive emotions (when they feel they are making progress) to negative emotions (when they feel they are blocked), and then back to positive emotions again. In each direction, the magnitude may be relatively large. If we could graph the emotional state of the problem solver against time, the graph for many students would resemble a periodic function.

Although a student's emotional state during problem-solving activities is expected to vary from positive to negative, it is reasonable to assume that a student whose experiences were consistently negative could develop a negative view of problem solving that would be stable and permanent. In terms of Spielberger's state-trait conception of anxiety (Spielberger, O'Neil, & Hansen, 1972), the emotional state could, through repetition, become a trait, with the result that the student would always be anxious when confronted with a problem-solving task. Buxton (1981) argues that this type of long-term negative reaction to mathematics is quite common, and he refers to it as the panic response. It seems reasonable to hope that a carefully constructed curriculum in mathematical problem solving could reduce the frequency of the panic response among students.

The importance of time in dealing with emotion has been noted by other writers. Kagan (1978) discusses both the duration of the emotional state and the rapidity with which the emotional state has an effect on the individual (the "rise time"). Wertime (1979) discusses the notion of "courage spans" in problem solving, where the courage span is the length of time that a student will persist in trying to solve a nonroutine problem. The courage span, like the attention span, does increase with age, but it can be reduced to almost nothing for students who are unable to deal with the frustrations of problem solving.

The relatively short duration of emotional states in problem solving can be contrasted with the typical conception of attitude toward mathematics (Haladyna, Shaughnessy, & Shaughnessy, 1983). Attitude is usually defined as a general, long-term emotional disposition toward the subject; it is supposed to be stable, not periodic. Also, attitude is not typically conceived as a very intense emotional state. In the terms used in this chapter, attitude is generally small in magnitude (although its direction may be either positive or negative).

## LEVEL OF AWARENESS

Problem solvers are frequently unaware of the emotions that are influencing them and their problem-solving processes. This lack of awareness is closely related to the notion of limited processing capacity discussed by Mandler (1984). Although an emotional reaction may come to consciousness, an awareness of its cause may not stay in consciousness very long; for example, an interruption in a problem solver's plan may cause frustration, and the solver may immediately reduce that frustration by giving up on the problem and setting a new goal or making a new plan that will not be interrupted. In observing students, it seems that this reduction of frustration happens very quickly, even automatically.

If we can help problem solvers become aware of their emotional reactions, they should improve their ability to control automatic responses or unconscious responses to problems. Because the automatic responses are frequently inadequate, greater awareness should improve the problem solver's chances for success.

## LEVEL OF CONTROL

Students may have more difficulty in controlling some emotions than others. Students with a deep-seated fear of mathematical problem solving, for example, may have difficulty in bringing that fear under control. Recent cognitive approaches to these extreme anxieties (Beck & Emery, 1985; Meichenbaum, 1977) suggest promising techniques that could be adapted for use in mathematics education; however, many of the more typical emotional reactions to problem solving may be relatively easy to control. Once students understand that problem solving involves interruptions and blockages, they should be able to view their frustrations as a normal part of problem solving, not as a sign that they should quit. Similarly, students should be able to learn that the joy of finding a solution to a problem is not a signal to relax and go on to another task. Instead, students can learn to view it as a cue to look back at their solution and check for reasonableness, elegance, and alternate approaches.

Most instruction in problem solving suggests that students try to control their cognitive processes and choose strategies that may be more effective than trial and error or other, more random approaches to problems. In the same way, instruction on affective issues should help students control their emotional reactions to the frustrations and joys of problem solving.

## Relating Affect to Instructional Issues

When students are involved in instruction in problem solving, their affective states are influenced by a variety of factors. Some of the factors that seem particularly important to mathematics include the different kinds of cognitive processes that the problems require, the setting within which the instruction takes place, and the belief systems that the student (and the teacher) hold.

### TYPES OF COGNITIVE PROCESSES

Affective reactions may have different influences on different types of cognitive processes. Metacognitive and managerial processes seem particularly susceptible to the influence of emotions; for example, decisions about whether to persevere along a possible solution path may be heavily influenced by student confidence or anxiety. Storage and retrieval processes may be less affected by the emotions; however, when the magnitude of the emotional response is large, reaching the panic level (Buxton, 1981), students indicate that all processing stops. All of working memory seems to be engaged in evaluating that emotional state.

Affective reactions may also be different at different stages of problem solving. In Polya's (1945) terms, the emotions at the looking-back stage will be very different from the frustrations of trying to carry out a plan of attack.

Causal attributions of success or failure (Fennema, 1982; Leder, 1986; Reyes, 1984) also appear to have a significant impact on metacognitive and managerial processes. Students who attribute their successes to help from the teacher, for example, may not feel capable of outlining alternative solution paths and then making reasonable decisions about which path to choose and why. There has been a substantial amount of research on the topic of attributions, but most of it has not been tied closely to the mainstream of research on mathematical problem solving. The possibilities for fruitful collaboration here seem worth pursuing.

### TYPES OF INSTRUCTIONAL ENVIRONMENTS

Problem-solving instruction can occur in a wide variety of instructional settings. Some of the most important of these variations are individual, small-group, or large-group instruction; the amount of direction provided by the teacher or materials; the use of instructional aids, such as computers, manipulatives, or other laboratory equipment; the kinds of assessment that the students expect; and many other variations. The different kinds of instructional environments present a very large category for investigation. Turkle (1984), for example, discusses children's interactions with computers and notes that these interactions generate very strong emotional responses. Children's beliefs about computers, as well as the tendency of some students to become completely engrossed in computer work, make the relationship of this instructional environment to affective influences particularly interesting.

The problem-solving environment is influenced by many factors that are not well understood. Why, for example, do teachers report that it is much more difficult to get a positive response from ninth-grade students than from seventh-grade students to instruction in problem solving? Why do students in a college-level calculus class often appear totally disinterested in nonroutine problems? The social forces that are at work in these different environments seem to play an important role.

### BELIEF SYSTEMS

Problem solving is supposed to be a part of life in mathematics classrooms; as a part of the classroom milieu, it is subject to social and anthropological forces that operate in that setting. As Mehan (1978) points out, ethnographic approaches are a useful tool for the analysis of life in classrooms. In research on problem solving, this point of view has generally focused on the student's belief systems and how these beliefs influence performance on mathematical problems.

Silver (1985) has argued that investigations of belief systems represent a promising new direction in research on problem solving, and Schoenfeld (1985) makes belief systems one of the cornerstones of his approach to studying problem solving. Both of these authors agree that belief systems lie on the border between affect and cognition.

Crossing borders frequently makes people uneasy, but some researchers are willing to cross over to the other side. D'Andrade (1981) discusses the role of emotion in cultural learning from an anthropological point of view, and he suggests that both beliefs and feelings are useful in memory and cognition. Rosamond (1982) deals with similar issues from a mathematics education perspective. The relationship between belief systems and affective factors in problem solving needs a broader, more anthropological approach than most researchers from mathematics education are prepared to take, and such an approach will take time to develop. If we are to make reasonable recommendations about problem-solving instruction in real classrooms, however, the relationship between affect and belief systems requires a deeper analysis than it has received so far.

## Summary and Implications

A theoretical framework for research on affect in mathematical problem solving should include the following factors:

1. Magnitude and direction of the emotion.
2. Duration of the emotion.
3. Level of awareness of the emotion.
4. Level of control of the emotion.

These factors constitute a reasonable beginning for an analysis of the characteristics of emotional states and their relationship to problem solving. Clarification of the role of these factors should have broad implications for both curriculum development and research.

In addition to these characteristics of the emotional states of problem solvers, we need to know much more about the ways in which these factors interact with the different types of cognitive processes, the different types of instructional environments, and the differing beliefs that students hold. A careful analysis of these issues should help reduce the current ambiguities that surround the area of affective influences on mathematical problem solving.

The main goal of research on affective issues in problem solving is the improvement of instruction and student performance. This research has special implications for those students who have weak problem-solving ability; for example, if we are going to expect all students (not just the most talented) to be able to solve nonroutine problems, then we should be prepared to deal with the emotional stress that this expectation is likely to produce, especially among the weaker students. Curriculum changes in the 1960s were not always ready for the cognitive limitations of younger mathematics students. Curriculum changes in the 1980s do not seem ready for the affective characteristics of our less successful problem solvers.

Although low-achieving students are of special concern, it seems clear that all mathematics students—even the strongest—need to be able to deal with the role of affect in their learning. Anecdotal evidence gathered from secondary school teachers suggests that some students who have gotten along well in mathematics in the past will suddenly lose all confidence when they have to address a more difficult problem than they are accustomed to. The transition to calculus, for example, is a traumatic experience for some of our more talented students; therefore, although research on affect is especially important for certain groups of students (especially those who are weak in problem solving), I have no doubt that a better understanding of affective issues in problem solving can be of benefit to the majority of students.

The current emphasis on problem solving in mathematics education has particular relevance for women and minorities, groups that have been underrepresented in mathematical careers. Data from National Assessment and a variety of other studies suggest that females do not perform as well as males on problem-solving tasks. Researchers have investigated a variety of factors that might explain these persistent differences. Some of the factors that have been considered are spatial ability (Fennema, 1982), patterns of enrollment in mathematics courses (Armstrong, 1981), and biological issues (Beckwith, 1983). None of these approaches has turned out to be very useful in explaining the differences in problem-solving achievement. Although it is likely that more than one factor is involved in these differences, it seems reasonable to expect that affective factors will be an important part of the explanation.

Achievement in problem solving by minority-group students also seems likely to be depressed by affective factors (Matthews, 1984). Other powerful influences (such as parental education or income level) probably play a significant role as well; however, it seems very likely that affective factors play a particularly important role in the achievement patterns for both women and minorities.

In conclusion, current efforts to improve problem solving are a major goal of mathematics education. For all students to achieve this goal, researchers and developers need to find better ways to address affective issues in research on problem solving. By integrating these new ideas on affect into current research and development efforts, we should be able to improve mathematical problem solving for all students, especially women and minorities.

*Acknowledgment.* Preparation of this paper was supported by the National Science Foundation (Grant No. MDR-8696142). Any opinions, conclusions, or recommendations are those of the author and do not necessarily reflect the views of the National Science Foundation.

## *References*

Adams, J.L. (1980). *Conceptual blockbusting: A guide to better ideas* (2nd ed.). New York: Norton.

Armstrong, J.M. (1981). Achievement and participation of women in mathematics: Results of two national surveys. *Journal for Research in Mathematics Education, 12*, 356–372.

Beck, A.T., & Emery, G. (1985). *Anxiety disorders and phobias: A cognitive perspective.* New York: Basic Books.

Beckwith, J. (1983). Gender and math performance: Does biology have implications for educational policy? *Journal of Education, 165*, 158–174.

Bloom, B.S., & Broder, L.J. (1950). *Problem-solving processes of college students.* Chicago: University of Chicago Press.

Bower, G.H. (1981). Mood and memory. *American Psychologist, 36*, 129–148.

Brown, A.L., Bransford, J.D., Ferrara, R.A., & Campione, J.C. (1983). Learning, remembering, and understanding. In J. Flavell & E. Markman (Eds.), *Mussen's handbook of child psychology* (Vol. 3, pp. 77–166). Somerset, NJ: Wiley.

Brown, S.I., & Walter, M. (1983). *The art of problem posing.* Philadelphia: Franklin Institute Press.

Burton, L. (1984). Mathematical thinking: The struggle for meaning. *Journal for Research in Mathematics Education, 15*, 35–49.

Buxton, L. (1981). *Do you panic about maths?* London: Heinemann.

Charles, R.I., & Lester, F.K., Jr. (1984). An evaluation of a process-oriented instructional program in mathematical problem solving in grades 5 and 7. *Journal for Research in Mathematics Education, 15*, 15–34.

D'Andrade, R.G. (1981). The cultural part of cognition. *Cognitive Science, 5*, 179–195.

Fennema, E. (1982, March). *The development of variables associated with sex differences in mathematics.* Symposium conducted at the annual meeting of the American Educational Research Association, New York.

Hadamard, J. (1945). *The psychology of invention in the mathematical field.* Princeton: Princeton University Press.

Haladyna, T., Shaughnessy, J., & Shaughnessy, J.M. (1983). A causal analysis of attitude toward mathematics. *Journal for Research in Mathematics Education*, *14*, 19–29.

Johnson-Laird, P.N. (1983). *Mental models: Towards a cognitive science of language, inference, and consciousness*. Cambridge, England: Cambridge University Press.

Kagan, J. (1978). On emotion and its development: A working paper. In M. Lewis & L.A. Rosenblum (Eds.), *The development of affect* (pp. 11–41). New York: Plenum Press.

Kantowski, M.G. (1977). Processes involved in mathematical problem solving. *Journal for Research in Mathematics Education*, *8*, 163–180.

Kantowski, M.G. (1980). Some thoughts on teaching for problem solving. In S. Krulik (Ed.), *Problem solving in school mathematics* (pp. 195–203). Reston, VA: National Council of Teachers of Mathematics.

Kilpatrick, J. (1985). Reflection and recursion. *Educational Studies in Mathematics*, *16*, 1–26.

Lawson, M.J. (1984). Being executive about metacognition. In J.A. Kirby (Ed.) *Cognitive strategies and educational performance* (pp. 89–109). Orlando, FL: Academic Press.

Leder, G.C. (1986, April). *Gender-linked differences in mathematics learning: Further explorations*. Paper presented at the Research Presession for the Annual Meeting of the National Council of Teachers of Mathematics, Washington, DC.

Lester, F.K., Jr. (1983). Trends and issues in mathematical problem-solving research. In R. Lesh & M. Landau, *Acquisition of mathematics concepts and processes* (pp. 229–261). New York: Academic Press.

Mandler, G. (1975). *Mind and emotion*. New York: Wiley.

Mandler, G. (1984). *Mind and body: Psychology of emotion and stress*. New York: Norton.

Mandler, G. (1985). *Cognitive psychology: An essay in cognitive science*. Hillsdale, NJ: Lawrence Erlbaum Associates.

Mason, J., Burton, L., & Stacey, K. (1982). *Thinking mathematically*. London: Addison-Wesley.

Matthews, W. (1984). Influences on the learning and participation of minorities in mathematics. *Journal for Research in Mathematics Education*, *15*, 84–95.

McLeod, D.B. (1985). Affective issues in research on teaching mathematical problem solving. In E.A. Silver (Ed.), *Teaching and learning mathematical problem solving: Multiple research perspectives* (pp. 267–279). Hillsdale, NJ: Lawrence Erlbaum Associates.

Mehan, H. (1978). Structuring school structure. *Harvard Educational Review*, *48*, 32–64.

Meichenbaum, D. (1977). *Cognitive behavior modification: An integrative approach*. New York: Plenum Press.

National Council of Teachers of Mathematics. (1980). *An agenda for action*. Reston, VA: Author.

Norman, D.A. (1981). Twelve issues for cognitive science. In D.A. Norman (Ed.), *Perspectives on cognitive science* (pp. 265–295). Norwood, NJ: Ablex.

Ohio Department of Education. (1980). *Problem solving: A basic mathematics goal*. Columbus: Author.

Polya, G. (1945). *How to solve it*. Princeton: Princeton University Press.

Resnick, L.B., & Ford, W.W. (1981). *The psychology of mathematics for instruction*. Hillsdale, NJ: Lawrence Erlbaum Associates.

Reyes, L.H. (1984). Affective variables and mathematics education. *Elementary School Journal*, *84*, 558–581.

Rigney, J.W. (1980). Cognitive learning strategies and dualities in information processing. In R.E. Snow, P.A. Federico, & W.E. Montague (Eds.), *Aptitude, learning, and instruction* (Vol. 1, pp. 315–343). Hillsdale, NJ: Lawrence Erlbaum Associates.

Rosamond, F. (1982). Listening to our students. *For the Learning of Mathematics, 3*, 6–11.

Schoenfeld, A.H. (1983). Beyond the purely cognitive: Belief systems, social cognitions, and metacognitions as driving forces in intellectual performance. *Cognitive Science, 7*, 329–363.

Schoenfeld, A.H. (1985). *Mathematical problem solving*. Orlando, FL: Academic Press.

Shiffrin, R.M., & Dumais, S.T. (1981). The development of automatism. In J.R. Anderson (Ed.), *Cognitive skills and their acquisition* (pp. 11–140). Hillsdale, NJ: Lawrence Erlbaum Associates.

Shufelt, G. (Ed.). (1983). *The agenda in action*. Reston, VA: National Council of Teachers of Mathematics.

Silver, E.A. (1982, January). *Thinking about problem solving: Toward an understanding of metacognitive aspects of mathematical problem solving*. Paper prepared for the Conference on Thinking, Fiji.

Silver, E.A. (1985). Research on teaching mathematical problem solving: Some underrepresented themes and needed directions. In E.A. Silver (Ed.), *Teaching and learning mathematical problem solving: Multiple research perspectives* (pp. 247–266). Hillsdale, NJ: Lawrence Erlbaum Associates.

Simon, H.A. (1982). Comments. In M.S. Clark & S.T. Fiske (Eds.), *Affect and cognition. The Seventeenth Annual Carnegie Symposium on Cognition* (pp. 333–342). Hillsdale, NJ: Lawrence Erlbaum Associates.

Spielberger, C.D., O'Neil, H.F., & Hansen, D.N. (1972). Anxiety, drive theory, and computer-assisted learning. In B.A. Maher (Ed.), *Progress in experimental personality research* (Vol. 6, pp. 109–148). New York: Academic Press.

Strongman, K.T. (1978). *The psychology of emotion* (2nd ed.). New York: Wiley.

Turkle, S. (1984). *The second self: Computers and the human spirit*. New York: Simon and Schuster.

Wertime, R. (1979). Students, problems, and "courage spans." In J. Lochhead & J. Clement (Eds.), *Cognitive process instruction* (pp. 191–199). Philadelphia: Franklin Institute Press.

Whimbey, A., & Lochhead, J. (1982). *Problem solving and comprehension* (3rd ed.). Philadelphia: Franklin Institute Press.

Wickelgren, W.A. (1974). *How to solve problems*. San Francisco: W.H. Freeman.

# 3
# Describing the Affective Domain: Saying What We Mean[1]

LAURIE E. HART[2]

For some time, I have been aware of problems with the definitions we use for what is typically called the *affective domain*. In discussions about the affective domain, psychologists, mathematics educators interested in research on problem solving, and mathematics educators interested in research on attitudes toward mathematics have had difficulty communicating clearly with one another owing to, in part, the lack of common usage of terms. These three groups of people seem to be using the same terms to mean different things and different terms to mean the same thing. The following quote from Herb Simon communicates some of my concerns.

> Turning now to "affect"...I encounter a...difficulty. Here I have some impression ...that we are indeed the traditional blind men, now touching one part of the elephant, now another. Affect is a word of everyday language that is subject to the imprecision of all such words—perhaps to more imprecision than most. Its various meanings are connected—that's how they arose in the first place—but not synonymous. *If our science is to advance, we must identify these nuances and both construct and adhere to a vocabulary that makes the necessary distinctions in a consistent way* [italics added] (Simon, 1982, p. 334).

The goals of this chapter are to describe the meanings different people ascribe to the words *attitude, affect, affective domain, belief system, emotion,* and *anxiety* and to summarize some of the consistencies and inconsistencies among the meanings.

## Background

Affective variables have been studied in connection with mathematics education in several ways. It is a goal of many mathematics educators that their students have positive attitudes toward mathematics. Some educators are convinced that positive attitudes will improve the ability of students to learn mathematics. Other educators view positive attitudes as an important educational outcome, regardless of the impact of attitudes on student learning. Another perspective on attitudes toward mathematics is that the best way to foster positive attitudes in students is to increase the level of understanding of mathematics. Some educators combine these perspectives in various ways.

In other words, some educators posit a causal connection between attitudes toward mathematics and achievement in mathematics with a change in attitude bringing about a corresponding change in achievement. Others posit a cause-effect connection between attitudes and achievement but think that achievement, rather than attitudes, is the causal factor. It may be that each of these hypotheses is partly true and that the relationship between attitudes and achievement in mathematics is reciprocal, with positive attitudes leading to greater learning and increased understanding leading to more positive attitudes. We do know that there are positive correlations between many affective variables and achievement in mathematics. To date, however, research does not give us a clear picture of general causal relationships between attitudes toward mathematics and achievement in mathematics.

It is relatively clear that decisions about how many and which mathematics courses to take in middle school, high school, and college can be influenced by affective characteristics of the student that have developed over a period of many years. The number of mathematics courses taken in high school affects the types of majors open to students in post-secondary education; this, in turn, affects career choice.

Certain groups of students have been identified that are not achieving their full potential in mathematics. Female students, minority students, and students from families with low socioeconomic status have not participated in mathematics and mathematics-related activities to the degree that their abilities predict. Affective variables have been found to be related to the underrepresentation of these groups in mathematics classrooms and careers requiring knowledge of mathematics.

## Definitions

Research on the affective domain in mathematics education is in need of a strong theoretical basis that will be developed only through sustained, systematic efforts over time. Too much of the research in this area has had no theoretical rationale. Often researchers include an affective component in a study with little thought or planning. Many studies have compared the relative effects of two instructional methods on student achievement and attitude in mathematics without including any careful review of the literature concerning attitudes. When this is the case, it is unlikely that important new knowledge will result.

In many studies, researchers dealing with a specific affective variable, such as mathematics anxiety, fail to examine the large bodies of literature from psychology on related topics, such as general anxiety and test anxiety. This has contributed to the difficulty of making clear definitions of the affective construct under study. Sometimes no description or definition of a particular variable is included in the research report. This makes interpretation of results difficult and detracts from efforts to compare results across studies. See Leder (1985) and Reyes (1984) for more details about affective variables and mathematics.

The terms described in the following sections vary in several dimensions, such as the amount of emotion involved and the importance of context for a particular construct. Anxiety would be considered highly emotional, whereas perceptions of the usefulness of mathematics would be much less emotional. Nonetheless, both are valid constructs with clear relationships to achievement in mathematics and election of optional courses in mathematics. We might designate anxiety as hot because of its emotional nature and perceived usefulness as cold because of its low emotionality.

Affective reactions often vary from one context to another. If a student is generally quite confident in mathematics, the student may be more confident about algebra than geometry and more confident when working alone than when explaining a concept to a class of peers. A student may not be anxious about mathematics in general, but may become anxious when trying to solve a nonroutine mathematics problem. A student typically bored with a mathematics class may become very interested when challenged to think about a nonroutine problem.

## Attitude

*Attitude* is generally defined by psychologists as a predisposition to respond in a favorable or unfavorable way with respect to a given object (i.e., person, activity, idea, etc.). This definition has three components: (a) an affective or emotional reaction to the object, (b) behavior toward the object, and (c) beliefs about the object (Rajecki, 1982). In other words, a positive or negative attitude toward mathematics could be inferred from one's emotional reaction to mathematics, one's behavior in approaching or avoiding mathematics, and one's beliefs about what mathematics is and how it may be used. Some authors, however, define attitude in a different way, including only beliefs about the object (e.g., Wyer, 1974). Such a definition eliminates the emotional and behavioral components of attitude and makes attitudes synonymous with beliefs.

Mathematics educators have used attitude in a less clearly defined way than the psychologists. During the 1960s and early 1970s, the interest of mathematics educators in the affective domain was usually limited to what was called "attitude toward mathematics." Most of the paper-and-pencil scales developed at this time measured the degree of one's liking or disliking for mathematics. Some of the instruments included items that dealt with both anxiety aroused by mathematics and liking for mathematics, making the meaning of the scores on the instruments difficult to interpret.

Later, the paper-and-pencil scales developed by mathematics educators show a multidimensional view of attitudes toward mathematics (e.g., scales from the National Longitudinal Study of Mathematical Abilities [Crosswhite, 1972], the Mathematics Attitude Inventory [Sandman, 1980], and the widely used

Fennema-Sherman Mathematics Attitude Scales [Fennema & Sherman, 1976]). These scales were designed to measure more specific components of attitudes and include clear descriptions of the constructs to be measured; for example, there now exist attitude scales designed to measure a student's perceptions of how useful mathematics is and will be for them in the future, and other scales to measure a student's self-concept or confidence with respect to mathematics.

An examination of the attitude instruments used by mathematics educators indicates that attitude means any one of a number of perceptions about mathematics, oneself, or one's mother, father, or teacher. These do not have a strong emotional component. Attitude has also been used by mathematics educators to mean anxiety, which typically does contain a strong emotional component. Many social psychologists would not include anxiety among the attitudes.

## Affect and Affective Domain

*Affect* has been used by psychologists in a variety of ways (Benner, 1985; Clark & Fiske, 1982; Corsini, 1984; Mandler, 1984). According to the *Encyclopedia of Psychology*, affect refers to "a wide range of concepts and phenomena including feelings, emotions, moods, motivation, and certain drives and instincts" (Corsini, 1984, p. 32). Using this definition, examples of affect would be anger, joy, fear, pride, hate, and anxiety. Affect is also sometimes used loosely as a synonym for feeling, emotion, and mood.

Simon (1982) distinguishes among several aspects of affect. He uses affect as a generic term and describes its main forms as "emotion," "mood," and "valuation." The emotion aspect is associated with the affect that can interrupt the attentional mechanism of the human nervous system and direct attention to a present danger or requirement. Surprise, fear, and anger are particular examples of this aspect of affect.

Mood, Simon's second aspect of the meaning of affect, is not as acute or as interruptive as emotion. Moods, such as sadness and happiness, are able to influence cognitive activities; for example, a particular mood can make us more likely to remember some events from the past and less likely to remember others.

People attach evaluations, Simon's third aspect of affect, to all sorts of things and events. Some of these evaluations are stored in long-term memory. These evaluations have two sources. According to Simon (1982): "Evaluations may be memories of affect associated with the object and event on some occasion or occasions when we experienced it. We may remember that we were sad, or happy, or frightened when some particular event or kind of event occurred, and the memory may or may not also reinvoke the original emotion" (p. 335). These evaluations may also come from learned associations – that is, reading is essential, but mathematics is not. The cognitive evaluations may have been initially caused by the affect, or the cognitive evaluations may be associated with and evoke the affect.

Affect has also been used, particularly among educators, as a blanket term to describe attitudes, appreciations, tastes and preferences, emotions, feelings, and values. This use of affect stems partly from the description of the affective domain by Krathwohl, Bloom, and Masia (1964) in *Taxonomy of Educational Objectives: The Affective Domain*. Krathwohl et al. conceptualized the affective domain on the dimension of internalization or intensity of a broad set of constructs but did not distinguish carefully among these constructs. Other educators have consciously decided to use *affective* as a general term for beliefs, attitudes, and emotions. Fennema, for example, uses the term *affective variables* because the early work on attitudes toward mathematics was not careful about the definition or measurement of attitudes, and she prefers to avoid the terms used at that time.

In summary, psychologists often use the term *affect* to indicate hot, gut-level emotional reactions. It is often used by educators to mean a wide variety of beliefs, attitudes, and emotions ranging from cold to hot.

## Beliefs and Belief Systems

Interest in beliefs and belief systems has come mainly from cognitive psychology, whereas most of the interest in attitudes and affect has come from social psychology. Much of the work on beliefs and belief systems by psychologists originated during the 1960s in the area of artificial intelligence. Colby defines belief as a judgment of the "credibility of a conceptualization" (1973, p. 253). Credibility of a conceptualization has to do with whether one accepts, rejects, or suspends judgment concerning a set of concepts and the interrelationships among those concepts. According to Colby (1973), "Beliefs are nonobservable theoretical entities postulated to account for certain observable relations in human behavior" (p. 254).

For some psychologists, attitude and belief are different constructs. Rajecki (1982), for example, includes beliefs about an object as one of three components of attitude; the other two components are an affective or evaluative component and intended behavior toward the object. Other psychologists define attitude and belief as the same psychological construct (e.g., Wyer, 1974), thus removing the evaluative and behavioral components from attitude.

Mathematics educators have done little research on student beliefs about mathematics. It is only recently, with the calls for research on belief systems by Silver (1985), Schoenfeld (1985), and others, that interest in student beliefs has increased. Silver (1985) includes beliefs in his list of the 10 underrepresented themes and needed directions for research on mathematical problem solving. He cites research from Lester and Garofalo (1982), Lesh (1983), Schoenfeld (1983), and the third mathematics assessment of the National Assessment of Educational Progress (Carpenter, Lindquist, Matthews, & Silver, 1983), as evidence of the influence of beliefs on problem-solving performance.

Schoenfeld (1985) describes belief systems as conceptions about the nature of mathematics, specifically the constitution of mathematical arguments. He found that the purely cognitive components of his framework for the analysis of mathematical behavior did a poor job of predicting the problem-solving processes of students. Students who had gained the necessary mathematical knowledge from their coursework were not able to make use of this knowledge in problem solving. Their failure to use learned mathematical knowledge did not come from having forgotten it, but from not perceiving the knowledge as being useful to them. Because the knowledge was not seen as useful, they did not attempt to use it.

Years of being trained to use mathematics that the students do not understand, and years of passively reproducing mathematical arguments handed to them by others, take their toll. Many students come to think of mathematical results as intact, preexisting truths that are passed on "from above." They come to think of mathematics as being beyond the scope of ordinary mortals like themselves. They learn to accept what they are taught at face value without attempting to understand it since such understanding would necessarily be beyond their ability. Moreover, the students come to believe that whatever they forget must be given up as lost forever; not being geniuses, they have no hope of (re-)discovering it on their own. Save for the lucky few who learn (or are taught) that things can be different, these students become the passive consumers of "black box" procedures. . . . Even when they get the right answers, there is some question as to how much of the mathematics they really understand (Schoenfeld, 1985, p. 373).

Schoenfeld is interested in students' beliefs about the nature of mathematics and the ways in which mathematics can be used. He is also interested in how these beliefs limit students' understanding of mathematics and their ability to solve mathematical problems.

Some research on the socialization of mathematics teachers has examined teachers' beliefs about the nature of mathematics (e.g., Thompson, 1985). These researchers have, in part, examined the degree of consistency between what teachers say they perceive mathematics to be and what teachers actually do in their mathematics classes. Also of interest is the relationship between teachers' beliefs about mathematics and student learning. Belief and belief system have been used in essentially the same way by psychologists and mathematics educators.

## Emotion

Mandler (1984) says that we have no adequate definition of *emotion*, even though both the psychological and philosophical examinations of emotion have long histories. Emotions, according to the *Encyclopedia of Psychology*, are high-energy states of mind that give rise to feelings (Corsini, 1984). The strength of the feelings may vary, and a direction is usually attached to these states of mind; thus, surprise, euphoria, anger, anxiety, and fear vary in intensity and may be categorized as either positive or negative. Mandler (1984) describes two major traditions in research on emotions—organic and mental. In the organic tradition, it is argued that physiological events (e.g., tight muscles, perspiration), rather than thoughts, are the source of emotions. In the mental tradition, emotions are seen as the result of thinking. Mandler theorizes that the source of emotions is both physical and mental.

Until very recently, emotion per se had received no attention in research on mathematics education. McLeod (1985) and Silver (1985) have begun to discuss the potential value of understanding the relationship between emotions and performance in mathematical problem solving. This important work is just beginning.

## Anxiety

Anxiety is one of the most studied emotions. According to Benner (1985), it "may be defined as a subjective feeling of tension, apprehension, and worry, set off by a particular combination of cognitive, emotional, physiological, and behavioral cues" (p. 65). Spielberger (1972) distinguished between two forms of anxiety—state anxiety and trait anxiety. State anxiety is an acute reaction to some perceived threat. It is of relatively short duration and is only experienced occasionally during the life of an individual. Trait anxiety is a more consistent, habitual emotional response to the events of life. Paper-and-pencil instruments have been used successfully to distinguish between state anxiety and trait anxiety (Spielberger, Gorsuch, & Lushene, 1970).

Mathematics anxiety has been examined mainly by educators and is viewed as a form of state anxiety that is aroused in situations perceived as involving the use of mathematics (Byrd, 1982). Research on mathematics anxiety done by mathematics educators has not, with some important exceptions, used careful definitions of the construct. It has sometimes been viewed as a negative feeling toward or dislike of mathematics. Other researchers have conceptualized mathematics anxiety as a fear of mathematics. Mathematics educators have included anxiety among attitudes toward mathematics. Many social psychologists would categorize anxiety as an emotion or affective response rather than an attitude. Anxiety is seen by psychologists as a hot reaction. The view of mathematics anxiety as an attitude indicates a cooler, less intense emotional reaction to mathematics than a view of mathematics anxiety as a strong gut-level response to specific mathematics experiences.

# Why Define These Words?

The opinion held by most mathematics educators has been that beliefs, attitudes, and emotions are identifiable characteristics of individuals that are related to scores on tests of mathematics achievement, can predict student decisions to enroll in optional mathematics courses, and can help explain differential mathematics achievement by race and gender. Beliefs, attitudes, and emotions have been examined via scores on paper-and-pencil instruments and occasionally via individual student interviews. This view of beliefs, attitudes, and emotions might be called a black-box approach as opposed to a cognitive approach. In a cognitive approach, the processes that underlie responses to paper-and-pencil instruments are examined. The type of information gained from a black-box approach differs from that received from the cognitive approach in the same way that information yielded by a standardized test of mathematics achievement differs from that gained by a transcript of student processes from a think-aloud interview during work on a mathematical problem-solving task. The time and effort required to collect and analyze the data obtained from the think-aloud interview are much greater than the time and effort required for the paper-and-pencil instrument, but the information gained is a richer reflection of the student.

Note that instead of referring to affective variables, affect, or attitudes toward mathematics, I am now using the terms *beliefs*, *attitudes*, and *emotions*, which, for me, lessens the confusion associated with the meanings of affective variables, affect, and attitudes toward mathematics. Here, belief is used as it has been used by Thompson (1985), Silver (1985), and Schoenfeld (1985) – to reflect certain types of judgments about a set of concepts. Like Rajecki (1982), I use attitude toward an object to refer to emotional reactions to the object, behavior toward the object, and beliefs about the object. Emotion is used here, as in Mandler's book (1984), to represent a hot gut-level reaction. Affect is used only as a synonym of emotion; many educators use this term in a more general sense. By distinguishing among these terms in this way, we may help improve our understanding of the constructs and research about the constructs.

It is not the intent of this chapter to convince mathematics educators to adopt the definitions used by psychologists. Mathematics education research differs in important ways from psychological research. I want to encourage mathematics educators to understand the ways in which psychologists and mathematics educators differ in their views of these constructs and then to clearly describe the definitions used in their research.

## Footnotes

[1]An earlier version of this paper was presented at the annual meeting of the Research Presession to the National Council of Teachers of Mathematics, Anaheim, CA, April 7, 1987.

[2]The author's name has changed from Laurie Hart Reyes to Laurie E. Hart.

## References

Benner, D.G. (Ed.). (1985). *Baker encyclopedia of psychology.* Grand Rapids, MI: Baker Book House.

Byrd, P.A. (1982). *A descriptive study of mathematics anxiety: Its nature and antecedents.* Unpublished doctoral dissertation, Indiana University, Bloomington.

Carpenter, T.P., Lindquist, M.M., Matthews, W., & Silver, E.A. (1983). Results of the Third NAEP Mathematics Assessment: Secondary school. *The Mathematics Teacher, 76,* 652–659.

Clark, M.S., & Fiske, S.T. (Eds.). (1982). *Affect and cognition.* Hillsdale. NJ: Lawrence Erlbaum Associates.

Colby, K.M. (1973). Simulations of belief systems. In R.C. Schank & K.M. Colby (Eds.), *Computer models of thought and language* (pp. 251–286). San Francisco: W.H. Freeman.

Corsini, R.J. (Ed.). (1984). *Encyclopedia of psychology* (Vol. 1). New York: Wiley.

Crosswhite, F.J. (1972). *Correlates of attitudes toward mathematics* (National Longitudinal Study of Mathematical Abilities, Report No. 20). Palo Alto, CA: Stanford University Press.

Fennema, E., & Sherman, J. (1976). Fennema-Sherman mathematics attitudes scales. *JSAS Catalogue of Selected Documents in Psychology, 6*, 31. (MS. No. 1225).

Krathwohl, D.R., Bloom, B.S., & Masia, B.B. (1964). *Taxonomy of educational objectives: Handbook 2. Affective domain.* New York: David McKay.

Leder, G.C. (1985). Measurement of attitude to mathematics. *For the Learning of Mathematics, 5* (3), 18–21, 34.

Lesh, R. (1983, April). *Modeling middle school students' modeling behaviors in applied mathematical problem solving.* Paper presented at the annual meeting of the American Educational Research Association, Montreal, Canada.

Lester, F., & Garofalo, J. (1982, March). *Metacognitive aspects of elementary school students' performance on arithmetic tasks.* Paper presented at the annual meeting of the American Educational Research Association, New York.

McLeod, D.B. (1985). Affective issues in research on teaching mathematical problem solving. In E.A. Silver (Ed.), *Teaching and learning mathematical problem solving: Multiple research perspectives* (pp. 267–279). Hillsdale, NJ: Lawrence Erlbaum Associates.

Mandler, G. (1984). *Mind and body.* New York: Norton.

Rajecki, D.W. (1982). *Attitudes: Themes and advances.* Sunderland, MA: Sinauer Associates.

Reyes, L.H. (1984). Affective variables and mathematics education. *Elementary School Journal, 84*, 558–581.

Sandman, R.S. (1980). The Mathematics Attitude Inventory: Instrument and user's manual. *Journal for Research in Mathematics Education, 11*, 148–149.

Schoenfeld, A.H. (1983). Episodes and executive decisions in mathematical problem-solving skills. In R. Lesh & M. Landau (Eds.), *Acquisition of mathematics concepts and processes* (pp. 345–395). New York: Academic Press.

Schoenfeld, A.H. (1985). *Mathematical problem solving.* Orlando, FL: Academic Press.

Silver, E.A. (1985). Research on teaching mathematical problem solving: Some under-represented themes and needed directions. In E.A. Silver (Ed.), *Teaching and learning mathematical problem solving: Multiple research perspectives* (pp. 247–266). Hillsdale, NJ: Lawrence Erlbaum Associates.

Simon, H.A. (1982). Comments. In M.S. Clark & S.T. Fiske (Eds.), *Affect and cognition* (pp. 333–342). Hillsdale, NJ: Lawrence Erlbaum Associates.

Spielberger, C.D. (1972). Conceptual and methodological issues in anxiety research. In C.D. Spielberger (Ed.), *Anxiety: Current trends in theory and research* (Vol. 2, pp. 481–493). New York: Academic Press.

Spielberger, C.D., Gorsuch, R., & Lushene, R. (1970). *STAI Manual.* Palo Alto, CA: Consulting Psychologists Press.

Thompson, A.G. (1985). Teachers' conceptions of mathematics and the teaching of problem solving. In E.A. Silver (Ed.), *Teaching and learning mathematical problem solving: Multiple research perspectives* (pp. 281–294). Hillsdale, NJ: Lawrence Erlbaum Associates.

Wyer, R.S., Jr. (1974). *Cognitive organization and change: An information processing approach.* Potomac, MD: Lawrence Erlbaum Associates.

# Part II Studies of Learning

# 4
# Affect in Schema Knowledge: Source and Impact

SANDRA P. MARSHALL

In this chapter, I discuss the role of affect in cognitive processing. The importance of affect in processing mathematical information is described in the context of solving arithmetic story problems. More specifically, I offer some ideas about the way affective responses to mathematical problem-solving situations influence the development, maintenance, and retrieval of information stored in human memory. I outline a model of human memory based upon schema knowledge structures, and I suggest how affective information may be stored within a schema. The chapter is concluded with a discussion of some affective responses to mathematics evidenced by students' comments as they worked with story problems.

## Affective Responses to Story Problems

The topic is introduced with descriptions of two affective responses to solving arithmetic story problems. Although the situations are hypothetical, they generally correspond to situations that have elicited the two responses in practice. The responses themselves are not hypothetical. I have observed them repeatedly in my own research and in anecdotes related to me by colleagues. The situations are useful in my later consideration of how affect influences cognitive processing and how it may be stored in human memory.

### The Emotional Response

Imagine the situation in which a child is learning to solve arithmetic story problems. The child is presented with a problem and is asked to find the solution. The story problem is a typical textbook problem that has key words such as *altogether* or *have left*. The child guesses that addition would be appropriate and carries out the addition algorithm successfully.

Suppose that the solution is correct, and suppose that the child then attempts to solve another problem but does not get the correct answer. There are many opportunities for error. Some of the words may have been misread, or key words may have been imagined to be present when they were not. An incorrect association between a key word and an arithmetic operation could have been drawn. The operation may have been correctly identified but was confused in the child's understanding with another operation. The operation may have been correctly chosen, but there was a bug in the algorithmic application. Finally, there may simply have been a careless slip or transcription error in the writing down of some of the numbers.

The child probably will not recognize what is wrong with the solution and will not understand why the answer to the previous problem was correct and the answer to the current one incorrect. The child believes that the same thing was done in both situations, with the result that the child's actions were correct on one occasion and incorrect on another. What will the child's response be?

Although a single error may not cause the child to experience emotion about problem solving, repeated episodes of this type may lead to a sense of frustration, a distrust of the child's own skills, or a general feeling of unease. The immediate feedback the child receives from parents, teachers, and peers may cause additional emotional reactions, such as embarrassment or shame. These emotions, like other features of the situation, will influence the child's performance now and will also be encoded in memory as part of the experience of problem solving.

## The Attitudinal Response

Picture this same child several years later in situations in which the child is required to solve story problems. Now, whenever the child is presented with a problem, the initial response is, "I don't like story problems." Mere recognition of the situation is sufficient to trigger the affective response. The child does not need to be engaged in solving the problem.

This is an attitudinal response. It differs from an emotional one in at least two ways. First, the attitudinal response comes from the activation of previously stored affective memories. The emotional response comes about as the reaction to emotion that arises during the situation. Second, the attitudinal response is typically dispassionate (i.e., cold, rather than hot). The emotional response may be an intense feeling accompanied in extreme cases by nausea, increased heart beat, or shaking hands. The attitudinal response, in contrast, does not usually activate observable physiological reactions. In this sense, it is cold. It may still have an impact upon the child's willingness to engage in problem solving and may be itself sufficiently strong to block the child's attempts to search memory for appropriate techniques to solve the problem.

These are not the only two affective responses that can be observed in mathematics situations, but they are the most common. In the remainder of this chapter, I describe a model of cognition that accounts for how these responses are encoded in memory with other situational features and how their retrieval may influence the retrieval and activation of many aspects of problem solving.

# The Organization of Human Memory into
# Schema Knowledge Structures

Human memory has been described under several organizing principles. Tulving (1972), for example, talks about episodic versus semantic knowledge. Anderson (1983) focuses on procedural versus declarative memories. Hinton, McClelland, and Rumelhart (1986) describe microfeatures in models of parallel distributed processing. What is common in these hypothesized knowledge structures is the general organization of long-term memory into networks. Individual pieces of knowledge are viewed as nodes in the networks. These nodes may be linked together, or they may exist as isolates. Retrieval of information from memory depends upon where the information resides within the network. Information that has many links to other nodes usually has a higher probability of retrieval than knowledge that is unlinked because there are more paths through the linked network. Access to an isolated node demands a retrieval path directly to the node itself. Access to a highly connected node may be indirect, beginning with a node far removed from the target but connected to it over one or more paths through the network. In the latter case, access could begin with any of the connected nodes, rather than only with the target node.

The network concept of human memory helps us to understand some of the research findings in studies of expert and novice performance. Experts appear to have rich, highly interconnected networks. Novices are more likely to have fragmented, partially linked networks, possibly with inappropriate links between nodes. Psychological studies of retrieval and forgetting also support the network structure of memory. Comparisons of recognition and recall demonstrate that it is easier to retrieve knowledge from memory given cues that allow multiple paths through memory nodes than from cues that lead only to a single isolated node.

During the process of retrieval, how are some nodes selected and others ignored? Psychologists suggest that the links between nodes carry measured impulses, either positive or negative; thus, the *activation* of one node (perhaps through direct access) causes the activation of other surrounding nodes that are linked to the target. At the same time, this activation may also inhibit another set of nodes through negative connections.

It is a reasonable supposition that links between nodes vary in intensity. The degree to which activation spreads among nodes will be influenced by the strength of the associations that connect the nodes. Consequently, some nodes will receive a high degree of stimulation, whereas others will receive a lesser amount.

Through learning, both intentional and incidental, individual nodes are added to long-term memory, and groups of them become connected. I hypothesize here and elsewhere that the primary mechanism under which these connections are made and by which meaningful learning occurs is the *schema*. A schema is a knowledge structure that allows the individual to recognize aspects of his or her environment and to operate on them, either abstractly or concretely—that is, the schema governs the individual's interactions with the environment. Schemas are especially important in problem-solving situations because these situations demand responses of the individual, and schemas are the means by which these responses are constructed by the individual.

What constitutes a schema? In earlier research, I have developed a model of the schema built upon four basic components (see Marshall, Pribe, & Smith, 1987). First, there is a generic representation of the situation to which the schema applies. This component contains all of the facts, descriptors, and embellishments about the general instance in which the schema will be used. Related to this is the second component, which consists of the restrictions and conditions that must be met if the schema is actually to be instantiated. The first component, therefore, has the general description, and the second has the tests of goodness of fit of the description to the current situation. The third component contains planning mechanisms related to implementing the schema. Within this component are particular goals and subgoals that may be expected, as well as general goal-forming procedures. Finally, the fourth component has the actions and procedures that govern the actual implementation of the schema.

In a fully developed schema, each of these four components would be a subset of interconnected nodes, with links running between components as well as within them. Initial access to the schema could be through any of the four subsets, resulting in activation of the entire set of nodes that define the schema; thus, when faced with a problem, an individual might first recognize the general form of the problem (component one), might notice initially the presence or absence of a particular constraint (component two), might focus upon the obvious goals and secondary goals that must be achieved prior to solution (component three), or might identify particular actions that would be appropriate to the situation (component four). Each of these would activate the others. The depth to which any component might be activated and accessed by the individual depends upon the complexity of the problem. Trivial problems require little cognitive processing. Difficult ones might involve access to many different schemas.

This conception of memory organization has been applied mainly to the acquisition and storage of knowledge as they relate to cognitive skills. In particular, I have developed the model with respect to the knowledge required to solve arithmetic story problems (Marshall, 1987; Marshall, Pribe, & Smith, 1987). It is my purpose here to extend the model to include affective components of problem solving.

## Affective Links in the Schema Model

There are at least two means by which affect can enter schema knowledge. The affective features of the situation may be learned at the same time that other features of the schema are encoded. Alternatively, the affective response to the situation may be developed after the schema is fully formed and be, in effect, a secondary encoding related to the preexisting schema. I will consider each of these separately.

## Simultaneous Encoding

Return to the first description of an affective response to solving an arithmetic story problem. While solving the first story problem, the child encodes in memory certain aspects of the situation. Because the child is learning to solve these problems, he or she does not already have a schema that will guide and structure a response to the situation. The process of making the decision to add and of carrying out the algorithm creates weak bonds among features of the problem situation, such as the key word *altogether*, the process of making the choice of operation, and the action of carrying out the computation. Furthermore, if the child's answer is correct, another node may be linked to the others, indicating that tasks such as these are not difficult and perhaps are even pleasurable.

As the child encounters the second problem and makes an error, the link between the positive affect node will be weakened, and a competing link will be formed with a negative affect node. Repeated failures will strengthen this link. Repeated successes weaken it and strengthen the positive one.

For a single problem, the child is unlikely to encode an affect node unless the situation is exceptionally threatening or rewarding; however, if the child continues to attempt to solve problems and continues to err, a node of negative affect will be encoded, strengthened, and linked to the problem-solving process. With repeated failures and frustrations, the affect node becomes stronger, and its links to other features of the problem situation also become stronger. Eventually, one predicts that the presentation of a story problem will evoke a strong negative reaction from the child because the schema itself has been created in the presence of the affective response. In this case, affect is a feature of the situation and has been encoded along with other features.

Encoded in this way, the affect node is multiconnected, with links to many other nodes in the schema. It is not an isolate that becomes activated alone. Just as with any other feature of the problem-solving situation, it will have stronger links to some nodes and weaker links to others.

Where can affect nodes reside in the schema structure—that is, to which other nodes will affect nodes be strongly connected? Because errors of solution can be the result of incomplete or inappropriate elements in any of the four components of schema knowledge, it seems reasonable that negative (and positive) affect nodes can also be found in any one of the four components. Consider, for example, an individual who may have developed an affective response to a particular type of algebra problem that was a source of difficulty in the past. When faced with a problem that begins, "Two trains leave New York at the same time . . ." he or she immediately experiences a negative response. In this case, the form of the problem is part of the schema knowledge (the first component of general description), and the affect is linked directly to the encoding of *motion problems*. An equally difficult *mixture problem* ("Seth has 10 more quarters than dimes . . . . ") may evoke no affect or even positive affect, depending upon the emotional aspects of previous problem-solving experiences.

## *Posterior Encoding*

There appear to be cases in which individuals develop schema knowledge structures with little or no apparent affect links. One can imagine a competent mathematics student entering a mathematics contest and experiencing a negative reaction for the first time while attempting to solve a particular problem. This student will probably already have a highly developed set of schemas and be able to access them readily. Depending upon the strength of the affective reaction to the current experience, the student may encode the negative affect in such a way that it links to the schema(s) as a whole; thus, if the contest involves calculus problems, the student may develop an immediate dislike of all problems requiring integration. In this case, the affect node becomes connected to all parts of the schema. The schema nodes are already tightly linked and have probably achieved a level of activation that makes the instantiation of the schema appear automatic. When an affect node attaches to an existing schema, it connects equally to all parts of the schema. This bonding is in contrast to simultaneous encodings, in which affect nodes are linked more strongly to the elements with which they were first associated.

It is reasonable that simultaneous and posterior encodings will lead to different outcomes for individuals solving mathematics problems. Part of the difference comes in the specific versus diffuse connections between affect nodes and other nodes. When affect is encoded at the same time that other features are encoded, the links are specific, leading from one node directly to another. They are also relatively localized, extending primarily to nodes within one subcomponent of the schema. In contrast, posterior encodings lead to links that are more diffuse because they are formed between the affect nodes and the schema itself.

One can surmise that it may be easier to change affective responses that were coded simultaneously than to alter posterior encodings. Because the simultaneous encodings result from specific instances, they have links to identifiable parts of the schema. If positive experiences can be created that link to these same parts, a tension can be generated between the positive and negative responses to the same features of the problem. It is nice to think that many positive experiences could sufficiently weaken the older negative bond to the degree that the positive links would be dominant. Whether or not this is true is an empirical question, and it is an important research issue yet to be addressed.

## Examples of Affective Responses

There is some evidence consistent with the hypothesis that affect is coded as described in the preceding sections. This evidence comes from students' responses to open-ended questions about their problem-solving strategies and techniques.

## Description of Data

Several years ago, I undertook a research project that necessitated interviews of approximately 100 sixth-grade children enrolled in two elementary schools. Each child was interviewed for approximately 1 hour. During this time, the child responded to a traditional paper-and-pencil test of 10 story problems and then discussed with the interviewer an additional 10 story problems. Most of the problems required two computations for solution and involved whole numbers or fractions.

The children were asked to solve the problems on the paper-and-pencil test. They were not asked to find solutions to the problems discussed in the interview; instead, they were asked to describe how they might solve these problems, to talk about making a plan to solve them, and to point out important information in them. A discussion of the students' success in solving the problems and an examination of the strategies they used are presented elsewhere (Marshall, 1982). In this chapter, I describe their manifest affective reactions to the interview.

The interviewer in this study was a soft-spoken young woman who relates well to children and interacts easily with them. She had worked with handicapped children and had also taught children with reading difficulties. It is evident from the audiotapes that she encouraged the children to verbalize their feelings as they solved the problems, although this verbalization was not an explicit objective of the original study.

## Affective Responses

The children's responses during the interview were recorded in brief notes by the interviewer and audiotaped. The purpose of the original study was to examine the strategies used by sixth graders as they solved a set of problems. Of interest here is that most of the children volunteered affective information as well as details about their strategies (or lack thereof). The affective statements were interspersed throughout the interviews.

### Positive Responses

Most of the affective comments were negative, but there were some notable exceptions. One child made the following response as she checked her answer to a problem requiring the use of fractions: "I was right ... it adds up ... this is fun!" She went on to comment about the relation between addition and multiplication, and she was very pleased to recognize and describe the connection. She enjoyed showing the interviewer how multiple additions would yield the same answer as a multiplication computation. These responses suggest that the affect is linked to the procedures she used in solving the problems. This is an example of affect within the fourth component of schema knowledge.

Other positive responses were less specific. One of the students responded very confidently to one problem ("It's easy") and somewhat less confidently to the next one ("This one's a little harder"). He later responded with enthusiasm to a third problem ("I don't know how to solve it, but I know the answer"). (He did have the correct answer.)

There are two different affective responses here. The first statements are examples of values given to the problems based only upon reading the items. The affect nodes here are attitudinal and are probably connected to the features encoded in the general descriptive component of the schema. The affective response was made prior to any attempt to solve the problem. The third response by the student is positive. Even though he was unable to describe the procedures, he was certain that he understood the problem and had the answer, and he was correct. How this affective information is encoded is unclear. The student's understanding of the problem and his solution may result from a highly automated and fully activated schema. Individuals are frequently unable to access specific features of automatic responses. The student's confidence implies that he understands the situation despite his inability to describe his solution strategy.

NEGATIVE RESPONSES

As might be expected from other studies of affect in mathematics, many of the students' responses were negative. Most of the responses could be classified as cold, reflecting attitudes about the situation. Nonetheless, there were some hot affective responses, as shown by the following example.

Interviewer: Any idea about how to solve this one?
Child:       I think. I think that she ... she counted 7 heads and 24 legs ... um ... I
             had something ... I think she counted ... okay ... 7 heads and
             twenty ... (pause, trails off) ... okay ... I think she ... okay ... she
             counted 7 heads and I think there were 14 um ... parakeets and ...
             (pause) ... 10 hamsters or whatever.
Interviewer: Okay, how did you get that?
Child:       I went ... she had 7 heads so ... oh, my heart is beating so
             fast ... (trails off)
Interviewer: Are you scared?
Child:       Yeah.
Interviewer: Why? This is okay. Just relax. It's all right. Okay? You're doing a tremendous job. You are doing very well.

Following this exchange, the child continued to talk about the problem and her solution to it. After solving two more problems, however, her hands began to shake. The interviewer ended the session and spent several more moments reassuring the child that she had performed well on the tasks.

The majority of the hot responses were less dramatic. Several students reacted to the situation with steadily rising voices. By the end of the interview, these students were giving inflections to all of their statements, indicating a lack of confidence in their responses.

Negative affect was demonstrated most frequently in statements reflecting either a dislike of the task ("I hate this") or a self-judgment of the child's ability ("I'm no good at this"). Responses of both types are consistent with stored affect linked to various schema components.

Few affective reactions could be attributed to the first component of schema knowledge described previously—that is, students did not seem to have reactions to the general situations described in the problems. I hypothesize that such reactions are more likely to arise in other mathematics situations, such as algebra or calculus. At the sixth grade, children do not recognize situation similarities and, thus, would have no strong grouping of nodes to reflect the general description of various situations (Marshall, 1987).

There was evidence of affective links to other components of schema knowledge, particularly to the planning component. One student, for example, routinely ended her comments about solving each problem with negative statements, including the following: "This is probably wrong," "That's probably wrong," "I'm doing terribly." Most of these comments seemed to refer to her choice of operation and were made after she described why she elected to use a particular arithmetic operation. She did not voice hesitation as she carried out the computations; thus, it is likely that the negative affect for this child is linked to the third component (the planning and goal-setting component), rather than the procedural component, through which the actions are actually carried out.

Other children also expressed negative affective reactions about choosing an operation. One child commented, "I'm good with fractions but not word problems with fractions." Presumably, this means that the child feels confident when told what operation to execute but is hesitant about choosing the operation when it is not specified.

Some responses seemed to indicate the presence of affective links with the constraints found in the second component of schema knowledge (constraints and conditions for using the schema). Several students expressed the belief, "They're trying to trick you," without specifying who *they* might be. These students had difficulty understanding the problems. One said, "Oh, I hate these problems. . . . Why can't they just put numbers? . . . I don't understand them. . . . I don't like these." Generally, statements such as these were followed by the student's pronouncement that he or she could not solve the problem and would like to move to a different problem. These responses did not include references to planning or goal-setting considerations—that is, they attach to the second component of schema knowledge (the recognition of constraints that govern use of the schema).

## Summary

The responses of these children provide evidence of both hot and cold (emotional and attitudinal) reactions. There were clear physiological indicators, such as shaking hands, raised voices, and the self-report of increased heart rate. There were also unemotional statements of dislike and inability.

Most of the affective responses were in accord with the model of schema acquisition and use outlined previously. A tentative conclusion is that these responses support the simultaneous encoding of affect and problem features. There were some global responses to problem solving, but most of the children mentioned specific aspects that caused the distress.

Finally, it is encouraging that at least some of the students volunteered positive affective reactions. When students felt that they understood a problem and its solution, they spoke confidently and enthusiastically about solving it. For this group of students, positive affect appeared to be associated with their own self-evaluations of understanding.

*Acknowledgments*. Much of the theoretical research described in this chapter was supported by the Office of Naval Research under Contract No. N00014-85-K-0661. The empirical study from which the students' remarks are taken was funded by the National Institute of Education (Grant No. NIE-G-80-0095).

## References

Anderson, J.R. (1983). *The architecture of cognition*. Cambridge, MA: Harvard University Press.

Hinton, G.E., McClelland, J.L., & Rumelhart, D.E. (1986). Distributed representations. In D.E. Rumelhart & J.L. McClelland (Eds.), *Parallel distributed processing: Explorations in the microstructure of cognition* (Vol. 1, pp. 77–109). Cambridge, MA: MIT Press.

Marshall, S.P. (1982). *Sex differences in solving story problems: A study of strategies and cognitive processes* (Final Report, Grant No. NIE–G–80–0095). Washington, DC: National Institute of Education.

Marshall, S.P. (1987, April). *Knowledge representation and errors problem solving: Identifying misconceptions*. Paper presented at the annual meeting of the American Educational Research Association, Washington, DC.

Marshall, S.P., Pribe, C.A., & Smith, J.D. (1987). *Schema knowledge structures for representing and understanding arithmetic story problems* (Tech. Rep. Contract No. N00014–85–K–0661). Arlington, VA: Office of Naval Research.

Tulving, E. (1972). Episodic and semantic memory. In E. Tulving & W. Donaldson (Eds)., *Organization and memory* (pp. 382–403). New York: Academic Press.

# 5
# Aesthetic Influences on Expert Mathematical Problem Solving

EDWARD A. SILVER and WENDY METZGER

In the past two decades, considerable progress has been made in understanding the mechanisms underlying successful human problem solving. One valuable line of inquiry has focused on the differences between experts and novices in a variety of task domains, including mathematics and science (e.g., Frederiksen, 1984; Silver & Marshall, in press). This research has highlighted a number of features that distinguish expert from novice behavior. It has been established, for example, that experts have extensive and powerfully organized stores of domain-specific knowledge that they can access in solving problems in their discipline. Moreover, experts have flexible representation systems available, and they often engage in qualitative analyses of problems before beginning quantitative actions.

Research on the differences between experts and novices is valuable not only because it offers powerful models of successful problem solving and helps to explain problem-solving failures, but also because it makes salient some important tendencies and competencies that characterize expertise in a task domain and that might otherwise remain covert.

In this chapter, the importance of aesthetic judgment as a component of expert mathematical problem solving is explored. We argue that aesthetic judgments, which have heretofore received little attention in the research literature, may be important factors in expert mathematical behavior. In particular, we argue that experts do monitor and evaluate their problem-solving activity and that their monitoring and evaluation often has an aesthetic character. Furthermore, we identify some basic aesthetic principles that appear to underlie expert judgments. We also claim that expert judgments based on aesthetic principles sometimes have affective and emotional connections.

The chapter begins with a brief survey of various perspectives on the importance of aesthetic judgments; in particular, the views of Hadamard, Poincaré, and Krutetskii are considered. In the next section of this chapter, we consider some data, generally in the form of excerpts from interview protocols or summaries of those protocols, that illustrate the influence and role of aesthetic judgments on expert mathematical problem solving. The chapter concludes with a few speculations on the implications of this work.

# Aesthetic Factors in Mathematical Problem Solving

Although aesthetics has not generally received much explicit attention in the problem-solving research literature of the past two decades, a few mathematicians and psychologists have suggested various ways in which aesthetics may manifest itself during mathematical activity. Kline (1962), for example, suggested that aesthetics is one of several elements that influences the direction of mathematical activity. In Kline's view, the other influences include practical, scientific, philosophical, and artistic considerations as well as the desire for intellectual challenge. More recently, Davis and Hersh (1981), in their popular account of the nature of mathematics, gave considerable attention to the aesthetic character of professional mathematical activity.

Earlier in this century both Poincaré (1946) and Hadamard (1945) suggested that aesthetic emotion is a guide during mathematical discovery that enables mathematicians to make choices (both conscious and unconscious) about what directions to take. Krutetskii (1976) suggested that the aesthetic evaluation of the elegance of a problem solution is a common behavior among mathematically capable students. A common theme found in these observations is that aesthetic principles can serve as a filter through which one's mathematical activities must pass. In this section, we examine in somewhat more detail the perspectives of Poincaré, Hadamard, and Krutetskii.

## *Poincaré: The Aesthetic Emotion*

Henri Poincaré (1946), himself a first-class mathematician, was interested in the process of mathematical creation or invention. He asserted that mathematical invention takes place by choosing among possible combinations of ideas. In his view, the simple generation of combinations is necessary, but not sufficient, for invention to occur; invention involves choosing the useful combinations out of the multitude of possible combinations. In Poincaré's view, the generation of potentially fruitful ideas was the result of conscious work, but the final choice of combinations was unconscious and not purely intellectual; rather, it was based on aesthetic feeling, "the feeling of mathematical beauty, of the harmony of number and form, of geometric elegance" (p. 391). Only if a combination of ideas appealed to one's aesthetic feeling could the combination move into the consciousness as a sudden illumination; thus, aesthetics may serve as the basis for choice in mathematical invention.

Poincaré argued that aesthetic sensibility is a characteristic that distinguishes mathematicians from nonmathematicians. Poincaré suggested that sometimes when a sudden illumination occurs to a mathematician, it turns out to be incorrect. Although the assertion or result may subsequently be shown to be incorrect, the idea is often one that the mathematician would have liked to have been true for aesthetic reasons. As Poincaré suggests, "We almost always notice that this false idea, had it been true, would have gratified our natural feeling for mathematical elegance" (p. 392).

## *Hadamard: Unconscious and Conscious Aesthetic Choice*

Jacques Hadamard (1945) was also interested in the role of aesthetic feeling in mathematical invention. Hadamard agreed with Poincaré that important components of invention occurred in the unconscious and had an aesthetic character. He asserted that conscious work, followed by an intermediate period in which one is not consciously working on the problem, often results in a sudden inspiration. During this incubation period, the unconscious works on the problem, and when a combination is made that appeals to one's aesthetic sensibility, the idea moves into consciousness. Hadamard emphasized that the line between the conscious and unconscious is not strict and that ideas may move back and forth between them.

Although Hadamard used the terms *conscious* and *unconscious* to describe the process, we might recast his view in the language of modern cognitive psychology as one involving short-term and long-term memory, respectively. A person attends to information in short-term memory (conscious), and connections are made in long-term memory without attention to the process (unconscious). Interestingly, Hadamard disagreed with the use of the think-aloud research technique as a means of understanding the invention processes of incubation and illumination because the information in the unconscious during incubation may remain totally unknown to the subject and because the short amount of time the subject works does not allow enough time for incubation to occur. Hadamard's analysis and objections are almost identical to more recent information-processing analyses (e.g., Ericsson & Simon, 1984) that have confirmed the difficulty people have in accessing information that is no longer available in short-term or working memory.

According to Hadamard (1945), one's aesthetic sensibility drives not only the unconscious choices that lead to mathematical discovery, but also the more general choices about which direction of investigation to pursue: "The guide we must confide in is that sense of scientific beauty, that special esthetic sensibility, the importance of which he [Poincaré] has pointed out" (p. 127). Mathematicians may make choices of what direction to pursue in an investigation without any concrete knowledge of how fruitful the attempt may be, but they have "feelings" for which directions are likely to be useful. In Hadamard's opinion, these feelings pertain to beauty.

That sense of beauty can inform us and I cannot see anything else allowing us to foresee. At least contesting that would seem to me to be a mere question of words. Without knowing anything further, we *feel* that such a direction of investigation is worth following: we feel that the question *in itself* deserves interest, that its solution will be of some value for science whether it permits further applications or not. Everybody is free to call or not to call that a feeling of beauty (p. 127).

Considered broadly, Hadamard's view might suggest that any choice made while solving a problem could be an example of aesthetic sensibility at work. It is probably more reasonable and productive to recognize that the influence of aesthetics is somewhat more narrow. Some choices are made without particular aesthetic influence but with more practical considerations in mind. These considerations include whether a particular direction is likely to lead to the problem goal; thus, aesthetic or other personal goals interact with the problem goals in influencing the problem solver's behavior.

## Krutetskii: Striving for Elegance

V.A. Krutetskii (1976), an educational psychologist in the Soviet Union, studied the nature of mathematical activity, especially as it was manifested in the behavior of precocious children. He asserted that "distinctive aesthetic feelings have great value in mathematical activity" (p. 347). In his view, aesthetics has its greatest impact in the evaluation of solutions that have already been found. He found that the expression of aesthetic feelings was characteristic of the capable students he observed, particularly with regard to the elegance of a solution.

And their whole demeanor testified to the aesthetic feeling they were experiencing: their eyes sparkled, they rubbed their hands in satisfaction and smiled, they invited one another to admire a keen train of thought or a particularly elegant solution (p. 347).

The features of elegance that Krutetskii observed in the work of mathematically capable students were clarity, simplicity, and economy. According to Krutetskii, "Very typical of capable students is a striving for the most rational solution to a problem, a search for the cleanest, simplest, shortest and thus most 'elegant' path to the goal" (p. 284). On the other hand, Krutetskii also noted that average students did not share this concern for elegance and paid little attention to the quality of their solutions. Krutetskii's assessment of the particular aesthetic features of the mathematical thinking of very capable students is in accordance with Poincaré's assertion that mathematicians have some aesthetic sensibility that nonmathematicians lack. As Krutetskii noted, "A striving for simplicity and elegance of methods characterizes the mathematical thought of all prominent mathematicians, past and present" (p. 284).

Finally, Krutetskii observed not only positive aesthetic reactions to elegant solutions, but also negative reactions to inelegant ones: "And, on the other hand, many spoke directly of a feeling of dissatisfaction and annoyance when the solution they had found was 'crude', unwieldy or complicated, but they could not find a better one" (p. 286). Thus, Krutetskii's findings demonstrate that the emotional side of aesthetic evaluation of a solution can be either positive or negative.

## The Role of Aesthetics in Problem Solving

Based on the preceding discussion, we might expect to see two roles for aesthetics in the problem-solving behavior of expert problem solvers. The first is the guidance of decision making during problem solving, as suggested by Poincaré and Hadamard. The second is the evaluation of the elegance of a completed solution, as suggested by Krutetskii. Both of these roles can be classified as types of aesthetic monitoring and evaluation. Because other types of monitoring may also be significant, such as evaluation of how close one is to the problem goal, we would not expect aesthetic monitoring to appear all the time or to be the sole basis for assessment of problem-solving activites.

One important feature of aesthetic monitoring is its strong affective component. Poincaré used the term *aesthetic emotion*, and Krutetskii wrote of feelings of satisfaction or dissatisfaction based on the elegance of a solution. If an aesthetic assessment is positive, then the associated feelings are positive; if aesthetic assessment is negative, then the associated feelings are negative. These feelings may be followed by appropriate action. A feeling of dissatisfaction, for example, may result in the search for a more elegant solution. Thus, another interesting feature of aesthetic monitoring is that it may provide a clear link between the solver's metacognitive activies (e.g., monitoring and evaluating) and the solver's emotional response in a problem-solution episode.

# A Study of Aesthetic Factors in Mathematical Problem Solving

The data discussed in this section of the chapter were obtained as part of a more general study of mathematical expertise conducted by the second author (W.M.) with supervision from the first (E.S.). The purpose of the larger study was to examine the nature and character of mathematical expertise at two levels: graduate students pursuing advanced degrees in mathematics and professional mathematicians. For the purposes of this chapter, we focus on those aspects of the research that appear to be directly related to aesthetic factors.

## Subjects

Five mathematics professors at a large state university and three graduate students in mathematics were observed while solving problems. Each of the professors was actively engaged in research in his or her field of mathematical specialization. Two of the mathematicians were active in the field of linear algebra, one in abstract algebra, one in applied mathematics, and the other in theoretical computer science. Two of the graduate students were in a doctoral program in mathematics, and one was working toward a master's degree in applied mathematics.

## Procedure

Each subject was interviewed twice. During each interview, the subjects solved two problems and answered a series of general questions designed to provide a framework that might be used to interpret their problem-solving behavior.

The format of each interview was the same. Each interview was conducted on an individual basis with each subject, and each interview was audiotaped. An interview session lasted approximately 1 hour and consisted of three parts: an initial problem, some background questions, and a second problem. A subject was given 20 minutes in which to work on a problem while the interviewer observed. Subjects were asked to think aloud while they solved the problem, and they were prompted to verbalize whenever there were long periods of silence during the session. Following the 20-minute solution time, the interviewer asked follow-up questions to clarify aspects of and gain more information about the subject's behavior. Next, the interviewer asked some general questions about the subject's mathematical background, opinions and beliefs about mathematics, and problem solving. Finally, the subject was given a second problem to work on for 20 minutes, and follow-up questions were asked about this problem. If a subject finished a problem in a short time, he or she was given another problem to solve; thus, each subject solved a total of four or five problems during the two interviews. The data used in analysis consisted of the audiotaped interview protocols, the interviewer's notes, and the subject's written work.

## Problems

Problems were considered for inclusion in the study if they were nontrivial but had solutions that required knowledge well within the range of general knowledge of the subjects. Problems that were classical in nature or that had well-known canonical solutions were avoided. Of the 10 problems chosen for the study, 5 were geometry problems and 5 were algebra or number theory problems. The geometry problems were chosen to provide some basis for comparison with previous research. Two of the geometry problems came directly from those used by Schoenfeld (1985). The algebra and number theory problems were chosen to provide some variety within the discipline of mathematics, because previous research has tended to focus primarily on geometry problems. The problems appear in the Appendix at the end of this chapter.

The problem concerning the product of four consecutive integers (Problem 1, Appendix) was given to every subject during the first interview. Each of the other problems was solved by at least three subjects, including at least one professor and at least one graduate student.

## Results

Examples of aesthetic monitoring were found in the protocols obtained from all of the subjects in this study, and they were found most extensively in the problem-solving protocols generated by three of the professors. In our discussion of the results, we rely heavily on excerpts from the protocols of each of these three professors to illustrate some general observations about the aesthetic influences that were evident in the problem-solving behavior of these experts.

In particular, we discuss two different functional roles for aesthetics that were evident in their expert problem solving. In the first functional role, aesthetic principles provide the basis for either an evaluation of the elegance of results after a solution has been obtained or an appreciation of a problem before its solution. A second functional role for aesthetic principles is to act as a guide for decision making during problem solving and, consequently, to affect the path of a solution.

In addition to discussing the two functional roles of aesthetic principles, we also present some evidence that aesthetics appears to serve as a basis for linking metacognitive activity, such as monitoring and evaluation, and emotional response in problem solving.

## AESTHETIC EVALUATION OF ELEGANCE

Professor C's behavior best illustrates the aesthetic evaluation of elegance of results and solutions. In the four-consecutive-integers problem, Professor C generated many ideas, suggesting several directions that he did not pursue but might have pursued if he had had more time. The problem itself appealed to him. In the post-interview he remarked, "It is an interesting problem, actually, to fuss with this . . . . It is nice that you have all these simple little things and you can make calculations and then the question is: What can you make of it?" Perhaps this problem appealed to Professor C's aesthetic sensitivity to novelty, symmetry and patterns, and simplicity. He may have been referring to the fact that one can perform seemingly simple activities and gain deeper understanding from them, or perhaps he was referring to the apparent simplicity of the question, which can actually be treated in more depth than it first appeared. For Professor C (and some of the other experts in the study), there apparently was beauty in such a situation—one that juxtaposed simplicity and complexity.

Another problem in which Professor C clearly expressed an aesthetic response was the 3-4-5 triangle problem (Problem 9, Appendix). When Professor C first saw this problem, he smiled and said he found it strange. When asked about this reaction, he said: "This was really a problem that I had never seen before. Maybe that was part of it—as an appreciation of a problem that I certainly don't remember—something like that. Y'know, how can there be a new problem in geometry? Maybe because I've forgotten all the old problems." This appreciation appears to have an aesthetic component, and it may be related to Mandler's (1982) assertion that aesthetic response and value are closely linked to both novelty and familiarity.

Professor C worked diligently for 20 minutes on the 3-4-5 triangle problem, trying three or four different approaches, but he never obtained a satisfactory solution, in part because his equations turned out to be so messy. Afterwards, he asked the interviewer to show him a solution. The interviewer demonstrated a solution in which auxiliary lines were drawn from the vertices to the point P, thus subdividing the triangle into three smaller triangles with the original sides as bases and 3, 4, and 5 as the heights. Professor C described this solution as "beautiful" and "gorgeous" because one did not have to calculate the length of the side of the triangle. Professor C apparently judged this solution elegant because of its simplicity and ingenuity.

In general, these excerpts from Professor C's interview demonstrate the expression of aesthetic appreciation upon presentation of a problem, after being given a result, or after finding a result, rather than during the process of the solution. Professor C appeared not only to evaluate results but also to demonstrate an appreciation of the problems themselves. The aesthetic character of his actions appeared to be a response to the elegance or beauty (e.g., symmetry, simplicity, or ingenuity) of results, rather than a guide for choosing directions during the solution of the problem.

## AESTHETIC MONITORING AS A GUIDE IN DECISION MAKING

Aesthetic concerns appeared to play an especially important role in Professor D's problem-solving attempts, particularly in the problems relating to geometry. In his problem-solving episodes, it was fairly clear that Professor D was trying not only to solve the problems but also to fulfill aesthetic goals. In the inscribed triangle problem (Problem 5, Appendix), for example, Professor D stated that he knew he could solve the problem using calculus but that he wanted to avoid that, because calculus would give him messy equations. He solved the problem partly using a symmetry argument and partly using calculus. Still, he was surprised that the calculus actually worked. He was dissatisfied with his solution because it did not show him *why* the equilateral triangle was the desired triangle. He felt that there should be some geometric solution that revealed why an equilateral triangle has the largest area. The use of calculus failed to satisfy his personal goal of understanding, as well as his aesthetic desires for "harmony" between the elements of the problem and elegance of solution.

The use of algebra in the triangle construction problem (Problem 6, Appendix)' also troubled Professor D. He found the correct ratios algebraically, and he found that he could construct the proper lengths using auxiliary drawings and the Pythagorean Theorem. Nevertheless, he was dissatisfied and wanted to see a geometric rationale for the result:

Presumably when something is this simple, normally there's gonna be a neat little drawing that will show you the answer is what it is . . . . I feel that I would have to think about it to get what I considered a neat answer . . . . Just doing algebra and then saying, oh, well, so the answer's gotta be this probably doesn't fit within the genre or the style . . . . You might take a purist's point of view—that isn't the way to do this kind of geometry.

Professor D apparently sought coherence and harmony in his solution. He felt that he should not have to leave the domain of geometry to gain understanding of the problem, but rather that the structure of the problem should be coherent solely within the geometric framework. Apparently, he believed that not only should there be a geometric solution but that a geometric solution would be more elegant.

A different kind of example of aesthetic influence on cognitive actions in solving a problem was found in Professor C's attempts to solve the 10001 problem (Problem 3, Appendix). At one point during his solution, he considered the prime factorization of the numbers in the sequence. He noticed that the first number in the sequence factored into $137 \times 73$, and he remarked that this factorization was "wonderful with those patterns;" however, he also remarked that he did not think the information was going to be particularly useful in solving the problem. Nevertheless, he did try to use the pattern as a basis for further exploration of the problem, but he was unsuccessful. The fact that there was a pattern in the factorization appealed to him, and the aesthetic appeal was closely tied to cognitive actions that might have produced a successful solution. Unfortunately, Professor C's cognitive activities were partially flawed, and a successful solution was not obtained.

One may make two important observations about this last example. First, this example illustrates that mathematical beauty and functional utility are not equivalent for Professor C. Although he found the factorization pattern beautiful, he did not believe that it would be helpful. Second, this episode clearly demonstrates that aesthetic principles alone do not lead to *effective* cognitive actions. Although aesthetic principles may serve as a powerful guide for and influence on cognitive activities, they do not guarantee success.

AESTHETICS AND EMOTIONAL RESPONSE

Professor P appeared to base most of his monitoring and evaluation on factors other than aesthetic appeal. Nevertheless, there were a few interesting episodes in his solution protocols that suggest a possible link between aesthetic judgment and emotional response.

Professor P, who had a background in algebraic number theory, was quite interested in the question of whether the product in the $n$-consecutive-integers problem (Problem 4, Appendix) could be a perfect square, and asked several times if he had more time to work on it. One could attribute this interest to the aesthetic appeal the question held for him; however, it seems more likely that he knew the question was solvable. He stated, "I think that's known in the literature someplace, if I'm not mistaken." Perhaps this kind of question appealed to him because he chose to work professionally in number theory for part of his career and the question is of a type familiar to him (Mandler, 1982), or perhaps he chose to work in number theory because this kind of problem was attractive. He remarked, "This takes me back to my good old days when I worked in number theory."

Professor P's response to this problem may reveal an important link between emotion and aesthetics. His attraction to the problem and his positive emotional response appear to be genuine, and the interplay between emotion and aesthetics provided for a lengthy engagement with the problem.

Another incident from the interviews also provides a glimpse of the link between aesthetics and emotion. Professor P had little difficulty solving the matrix problem (Problem 10, Appendix). His approach was straightforward and problem specific. First, he tested cases, examining each possibility until he determined the largest number that could be the minimum. Then he verified his answer. At this point, he expressed disappointment in his method: "Rather sloppy way to do things, but I think it'll work . . . . I'm a little disappointed with this one 'cause I don't know an algorithm for doing that exactly." He had solved the problem, but he felt that there should be a more general method that could be applied. His striving for generality and simplicity appeared to be aesthetically motivated. He desired a solution that was more elegant—that is, more aesthetically pleasing—and he expressed disappointment, which we consider to be basically an emotional response.

Professor S provided another example of a possible link between aesthetic evaluation and emotional response. She considered the four-consecutive-integers problem quite unappealing and expressed no interest in working on the problem. In fact, she refused to attempt the problem and asked to work on a different problem from the set. For Professor S, integers are "things you count with." They hold no interest or beauty for her. She is, however, quite interested in manipulating and interpreting large amounts of real data, which are typically encountered in her field of numerical analysis. One might conjecture that making sense of this kind of data holds some beauty and fascination for her that she cannot find working in the more abstract realm of integers and number theory. Perhaps she is simply interested in those aspects of mathematics that are truly *applied* and have some real-world utility, and aesthetics may play a part in her interest. Aesthetic evaluation may play the role of guide or filter, providing a negative emotional reaction that leads her to abandon the problem.

### FUNCTIONAL RELATIONSHIPS

Although we have discussed the functional roles of aesthetics as if they were independent, there was clear evidence in the protocols that these functions are often intertwined and difficult to separate. In some instances, aesthetic monitoring appeared to be serving as both a guide for decision making and a filter through which one expressed appreciation or dislike for a problem or a solution. Consider Professor S's refusal to work on the four-consecutive-integers problem because it had no appeal for her, which suggests that her aesthetic evaluation served as a guide for her initial problem-solving decision; however, her aesthetically-based dislike of the problem was also an example of an emotional response to a problem, and the response had an important negative consequence on her willingness to engage the problem.

There were a few other instances of the interrelationship between the evaluative and guidance roles of aesthetic monitoring. One of the graduate students expressed a desire to do things in the "cleanest" way, thereby expressing one criterion for elegance: "You're afraid to do something because you feel it'll be so messy that you don't want to face it." Similarly, in Professor C's solution of the 3-4-5 triangle problem, all his ideas turned out to involve messy calculations. He suspected that there must be a better way—that is, simpler, with more clarity. His distaste for the "confusing mess" and desire for neat equations were the result of aesthetic as well as practical considerations. In both of these examples, the problem solver desired a neat, elegant solution. The aesthetic desire for elegance affected each problem solver's monitoring and evaluation behavior and led each problem solver to try a different way to solve the problem.

## Conclusions

Although our investigation was not explicitly aimed at uncovering and studying aesthetic influences on expert problem solving, aesthetic factors were quite salient in the protocols obtained from our expert mathematicians. In fact, aesthetic factors emerged from our analysis of the protocols as extremely important influences on the problem-solving behavior of the experts. Our general conclusions from this analysis are discussed in this section.

### THE IMPORTANCE OF AESTHETICS

The data from this study support the conclusion that mathematical problem-solving expertise is a function of taste as well as competence. Aesthetics appears to play two somewhat different functional roles in the problem-solving behavior of expert problem solvers: (a) aesthetics serves as a basis for post hoc evaluation of solutions or problems, and (b) aesthetic principles guide decision making during problem solving. The role of aesthetics in the post hoc evaluation of one's solution is an instance of the sort of aesthetic evaluation that Krutetskii identified and discussed with respect to mathematically precocious youth. The role of aesthetics as a guide in decision making while solving problems is closely tied to the kind of aesthetic judgment discussed by Poincaré and Hadamard.

There were fairly obvious individual differences among subjects in the tendency to express aesthetic judgments. Two factors appeared to affect the likelihood that aesthetic evaluations would appear in a subject's protocol: problem difficulty and think-aloud interference.

The interaction of problem difficulty and the solver's success in solving the problem appeared to influence the appearance of aesthetic valuations. In particular, when a subject was having difficulty with a problem, he or she was less likely to express any aesthetic feeling. We assert that this phenomenon is due to the taxing cognitive load carried by an individual who is having difficulty solving a complex problem. Conversely, we assert that when an expert knew that he or she could solve a given problem, the expert had the cognitive luxury of aesthetic considerations. The fact that aesthetic appreciation of results or methods is often shown after, rather than during, a problem-solution episode suggests that one may be better able to attend to aesthetic considerations after one has the immediate and pressing cognitive demands under control.

A second factor affecting the likelihood that aesthetic monitoring would appear in a subject's protocol was a subject's sensitivity to the intrusiveness of the think-aloud procedure on the problem-solving process. Several subjects expressed discomfort with the methodology used in the interviews. Because much of the aesthetic monitoring may normally be covert, the necessity to talk aloud throughout the solution episode may have created an unnatural situation that suppressed the expression of aesthetic judgments. On the other hand, several subjects said that the procedure did not interfere at all with their problem solving. In fact, they asserted that they enjoyed the opportunity to make public the conversations they often hold privately when solving problems.

## AESTHETIC MONITORING AND SCHOENFELD'S MODEL

The roles of aesthetic monitoring and evaluation described previously can fit into Schoenfeld's (1985) episodic model of problem-solving behavior fairly readily. According to Schoenfeld, the episodes during problem solving include reading, analysis, exploration, planning, implementation, and verification, with transitions between episodes. In this study, aesthetic influences appeared most frequently during verification, when the problem solver evaluated the elegance of a solution, such as in Professor P's matrix problem or Professor C's 3-4-5 triangle problem. Aesthetic evaluation also occurred during reading or analysis episodes, when the problem solver demonstrated an appreciation or dislike for a problem. Aesthetic monitoring acted as a guide during analysis, exploration, or planning, as demonstrated in Professor D's geometry problems. When aesthetic monitoring guides decision making, it may be linked to transitions between episodes, such as when Professor C abandoned one "messy" approach and tried another. Thus, aesthetic monitoring can act as any other monitoring behavior might in Schoenfeld's model.

One question pertaining to Schoenfeld's model is the question of whether a particular assessment is local or global. Local assessments are limited in scope, whereas global assessments affect the solution path in some major way. The aesthetic monitoring and evaluation observed in this study seemed to consist primarily of global assessment. One example of local aesthetic evaluation is Professor C's appreciation of the pattern of the factors 137 and 73 in the 10001 problem. Examples of global aesthetic monitoring include Professor D's desire to find elegant geometric proofs and Professor C's desire to find less messy, more elegant methods for solving the 3-4-5 triangle problem.

The role of aesthetics in metacognitive activity needs further exploration. In particular, the relative importance of fundamental aesthetic principles, such as parsimony, simplicity, and clarity, needs to be carefully examined. The influence of each of these principles was evident in the problem-solving behavior of the experts in this study, but further research is needed to clarify the role and importance of these individual principles.

## AESTHETICS LINKING METACOGNITION AND EMOTION

One difference between aesthetic monitoring and other forms of cognitive monitoring, including those discussed by Schoenfeld, is that aesthetic monitoring is not strictly cognitive but appears to have a strong affective component; that is, decisions or evaluations based on aesthetic considerations are often made because the problem solver "feels" he or she should do so because he or she is satisfied or dissatisfied with a method or result. As we noted earlier in some of the protocol excerpts, there appears to be a link between aesthetic evaluation and emotional response.

On the basis of the protocols examined in this study, we conclude that aesthetics can serve as a link betweeen one's monitoring and evaluation behavior and one's emotional response. In this capacity, aesthetics provides a link between cognitive and metacognitive activity and emotion. As might be predicted by Mandler's (1982, 1985) theories of emotion and cognition, the cognitive/emotional connection was evident for problems or solutions that were either very novel or very familiar to the solver; the novelty or the familiarity apparently evoked an emotional response. Although it may seem somewhat confusing or contradictory to discuss novelty and familiarity as being similarly related to the cognitive/emotional connection, it is important to note that the experience of both familiarity and novelty is accompanied by interruption and cognitive discrepancy. In Mandler's theory, both interruption and cognitive discrepancy would lead to arousal—emotional response that is often quite intense and positive. The role of aesthetic principles in linking metacognitive activity and emotional response deserves further study.

## Summary

The problem-solving behavior of expert mathematical problem solvers appears to be influenced by aesthetic considerations. We have seen that there are several ways in which aesthetic principles can influence expert problem solving. In this chapter we have not dealt with the comparison of experts and novices with respect to the influence of aesthetics. On the basis of some pilot work done before this study was undertaken, it was clear that the positive influence of aesthetic monitoring was almost exclusively found in expert protocols; it was not detected in the protocols of our unskilled novices. To understand more completely the relationship between aesthetics and expertise, it would be interesting to extend the work of Krutetskii and examine the problem solving of skilled novices or precocious youngsters to see if aesthetic influences could be detected. There is some disagreement as to how widespread an appreciation of mathematical aesthetics might be found in the general population. Papert (1980) argued that the mathematical activity of nonexperts was heavily influenced by factors "that have at least as much claim to be called aesthetic as logical" (p. 197). On the other hand, von Glasersfeld (1985) has argued that it is a misconception of mathematics educators that the aesthetic characteristics of mathematics "should be as obvious and entrancing to every lay human as is the beauty of the sunrise." Clearly, there is more work to be done on this subject.

We believe that the appearance of a strong influence of aesthetics on problem-solving behavior may be a hallmark of entry into *mathematical culture* — a culture in which elegance, parsimony, symmetry, coherence, simplicity, beauty, and other similar attributes are highly valued. The mathematicians and graduate students in this study gave strong evidence of accepting aesthetic principles as a basis for their actions. If this cultural view is correct, then an important educational issue can be raised. Papert (1980) has noted that "If mathematical aesthetics gets any attention in the schools, it is as an epiphenomenon, an icing on the mathematical cake, rather than as the driving force which makes mathematical thinking function" (p. 192). If mathematical aesthetics is an important component of the culture of mathematical thinking, we need to explore both the wisdom and feasibility of transmitting some of the aesthetic aspects of that culture to students at an earlier point in their mathematical education, especially through experiences that allow them to appreciate and reflect upon fundamental aesthetic principles. As Dreyfus and Eisenberg (1986) have noted: "Developing an aesthetic appreciation for mathematics is not a major goal of school curricula. This is a tremendous mistake" (p. 9). We continue to make this mistake at great expense to our students and to the development of mathematical culture in our schools.

There is, however, another side to the consideration of the role of aesthetic principles in school mathematics. We would be wise to reflect on the negative educational consequences of the perversion of some of these aesthetic principles; for example, a pedantic overemphasis on parsimony, simplicity, or clarity has dampened the spirit and imagination of many students. We would argue that it is important to communicate the *quest* for such attributes. It is not really the results of the application of those principles that are at the heart of the matter for education; rather, it is the inculcation of those principles as guides to behavior and as components of one's belief and value system. As we have seen in this study, mathematical results do not emerge from the work of professional mathematicians in their final polished form. The original products of mathematical thought are shaped, refined, and polished by the aesthetic principles discussed in this chapter. Instead of showing our students only the finished products, we must involve them more actively in the processes of shaping, refining, and polishing. In this way, they may begin to experience more fully the world of mathematics — a world in which aesthetic principles play an important role.

*Acknowledgements.* The work described in this paper was supported in part by the National Science Foundation through Grant No. MDR-8696142 to Douglas B. McLeod. That support does not imply National Science Foundation endorsement of the ideas or opinions expressed in this paper.

## Appendix: Problems Used in the Study

1. *Four-consecutive-integers problem.* What can you say about the product of any four consecutive integers?
2. *abc problem.* Why are all numbers of the form abc,abc divisible by 13?
3. *10001 problem.* Prove that there are no prime numbers in the infinite sequence of integers 10001, 100010001, 1000100010001, . . . .
4. *n-consecutive-integers problem.* Prove that the product of $n$ successive integers is always divisible by $n$!

5. *Inscribed triangle problem.* Three points are chosen on the circumference of a circle of radius R, and the triangle containing them is drawn. What choice of points results in the triangle with the largest possible area? Justify your answer.

6. *Triangle construction problem.* You are given a fixed triangle T with base B, as in the figure below. Show that it is always possible to construct, with straightedge and compass, a straight line that is parallel to B and that divides T into two parts of equal area. Can you similarly divide T into five parts of equal area?

T

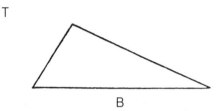

B

7. *Triangle trisection problem.* Trisect the area of a given triangle. That is, you should locate a point X inside the given $\triangle ABC$ so the $\triangle XBC$, $\triangle XCA$, and XAB are equal in area. Justify your answer.

8. *Square problem.* In the figure below, ABCD is a square, $\angle ECD = \angle EDC = 15°$. Show that triangle AEB is equilateral.

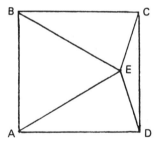

9. *3-4-5 triangle problem.* Consider the equilateral triangle below, where the point P is a perpendicular distance of 3, 4, and 5 from each of the sides, as shown. Show that the height of the triangle is 12. Can you generalize this result?

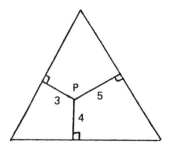

10.  *Matrix problem.* Given the following matrix of 25 elements choose five of these elements, no two coming from the same row or column, in such a way that the minimum of the five elements is as large as possible. Prove that your answer is correct.

$$
\begin{bmatrix}
11 & 17 & 25 & 19 & 16 \\
24 & 10 & 13 & 15 & 3 \\
12 & 5 & 14 & 2 & 18 \\
23 & 4 & 1 & 8 & 22 \\
6 & 20 & 7 & 21 & 9
\end{bmatrix}
$$

## References

Davis, P., & Hersh, R. (1981). *The mathematical experience*. Boston: Birkhäuser.

Dreyfus, T., & Eisenberg, T. (1986). On the aesthetics of mathematical thought. *For the Learning of Mathematics, 6*, 2–10.

Ericsson, K. A., & Simon, H. A. (1984). *Protocol analysis: Verbal reports as data*. Cambridge, MA: MIT Press.

Frederiksen, N. (1984). Implications of cognitive theory for instruction in problem solving. *Review of Educational Research, 54*, 363–408.

Hadamard, J. (1945). *The psychology of invention in the mathematical field*. Princeton: Princeton University Press.

Kline, M. (1962). *Mathematics: A cultural approach*. Reading, MA: Addison-Wesley.

Krutetskii, V.A. (1976). *The psychology of mathematical abilities in schoolchildren* (J. Teller, Trans.). Chicago: University of Chicago Press.

Mandler, G. (1982). The structure of value: Accounting for taste. In M.S. Clarke & S.T. Fiske (Eds.), *Affect and cognition* (pp. 3–36). Hillsdale, NJ: Lawrence Erlbaum Associates.

Mandler, G. (1985). *Cognitive psychology: An essay in cognitive science*. Hillsdale, NJ: Lawrence Erlbaum Associates.

Papert, S. (1980). *Mindstorms: Children, computers, and powerful ideas*. New York: Basic Books.

Poincaré, H. (1946). *The foundations of science* (G.B. Halsted, Trans.). Lancaster, PA: Science Press. (Original work published 1913.)

Schoenfeld, A.H. (1985). *Mathematical problem solving*. Orlando, FL: Academic Press.

Silver, E.A., & Marshall, S.P. (in press). Mathematical and scientific problem solving: Findings, issues and instructional implications. In B.F. Jones & L. Idol (Eds.), *Dimensions of thinking and cognitive instruction* (Vol. 1). Hillsdale, NJ: Lawrence Erlbaum Associates.

von Glasersfeld, E. (1985, July). *How could children not hate numbers?* Paper presented at the conference on the Theory of Mathematics Education, Bielefeld, Germany.

# 6
# Self-Confidence, Interest, Beliefs, and Metacognition: Key Influences on Problem-Solving Behavior[1]

FRANK K. LESTER, JOE GAROFALO,
and DIANA LAMBDIN KROLL

Any good mathematics teacher would be quick to point out that students' success or failure in solving a problem often is as much a matter of self-confidence, motivation, perseverance, and many other noncognitive traits, as the mathematical knowledge they possess. Nevertheless, it is safe to say that the overwhelming majority of problem-solving researchers have been content to restrict their investigations to cognitive aspects of performance. Such a restricted posture may be natural for psychologists and artificial intelligence scientists who are concerned primarily with expert systems or machine intelligence, but it simply will not suffice for the study of problem solving in school contexts.

In 1986 we began a study of the role of metacognition in the mathematical problem-solving behavior of seventh-grade students that was preceded by about 5 years of preliminary work in this area. Although we are convinced that metacognition plays a vital role in problem solving, we regard it as but one of a number of *driving forces* (to use the term coined by Silver [1982] and elaborated on by Schoenfeld [1985, 1987]). At least two other domains seem particularly important: *affects and attitudes* and *beliefs*. The purpose of this chapter is to present our ideas about the nature of the factors that affect success during problem solving. We begin with a discussion of four types of noncognitive and metacognitive factors. This is followed by several illustrations of the strong influence that certain factors (viz., self-confidence, interest, beliefs, and metacognition) can have on the problem-solving performance of seventh graders. The chapter concludes with a brief discussion of some conjectures related to the nature of the relation of these factors to problem solving.

## Theoretical Considerations

We begin by postulating that an individual's failure to solve a problem successfully when the individual possesses the necessary knowledge[2] stems from the presence of noncognitive and metacognitive factors that inhibit the appropriate utilization of this knowledge. These factors are of at least four types: affects and attitudes, beliefs, control, and contextual factors. In the following sections, we discuss each of these types.

## *Affects and Attitudes*

This domain involves individual feelings and includes emotions, preferences, and attitudes (cf. Hart, this volume, Chapter 3). Typically, emotions are subjective reactions to specific situations. Of course, emotions can have either a facilitating or debilitating effect on the individual, but negative emotions (e.g., frustration) are not necessarily debilitating, nor are positive emotions (e.g., joy) necessarily facilitating. There is a growing body of research to support the notion that emotions and cognitive actions interact in important ways (Mandler, this volume, Chapter 1). Most of this research has been restricted to the study of the conditions under which certain emotions occur or to the nature of an individual's behavior when in a particular emotional state. Because the primary focus of our research is metacognition, we have not investigated emotions and their effect on problem-solving performance. Moreover, emotions have not been given much attention in any other research studies of mathematical problem solving. Consequently, we cannot discuss emotions further. We do recognize the potential of emotions to influence problem-solving behavior, however.

Although some researchers interested in mathematical problem solving have focused their attention on studying preferences (e.g., likes and dislikes) and attitudes, research involving these constructs often has been limited to examinations of the correlation between preferences or attitudes and problem-solving performance (Silver, 1985). Not surprisingly, of the preferences and attitudes that have been studied with respect to problem solving, only confidence in learning and doing mathematics has been shown consistently to be positively correlated with achievement in mathematics (Reyes, 1984). In addition, Webb, Moses, and Kerr (1977) have suggested that tolerance of ambiguity and resistance to premature closure are also important correlates of problem-solving performance. Unfortunately, as McLeod (1985) has pointed out, the theoretical underpinnings of the research in this area have been very weak.

It is generally accepted that preferences and attitudes are traits, albeit perhaps transient ones, of the individual, whereas emotions and moods are situation- or time-specific states. An individual may have developed a particular attitude toward a class of problems that affects his or her performance on a specific problem in that class (e.g., a student may greatly dislike problems involving percents). At the same time, a particular problem may give rise to an unanticipated emotion (e.g., frustration may set in when a student, after working diligently on a problem for a considerable amount of time, recognizes that little progress has been made). At present, we are confining our attention to preferences and attitudes. More specifically, in this chapter we consider students' interest in problem solving (a preference) and confidence in their own mathematical problem-solving ability (an attitude). Interest in problem solving involves a liking for or willingness to engage in solving problems. Of course, the extent of an individual's willingness to solve problems is dependent upon the conditions under which the individual is working as well as the nature of the problems involved. Confidence includes the individual's belief in his or her ability to succeed in solving even hard problems as well as self-confidence with respect to his or her peers. Self-confidence actually is a consequence of a belief about self.

## Beliefs

Schoenfeld (1985) refers to beliefs (or *belief systems*, to use his term) as the individual's mathematical world view—that is, "the perspective with which one approaches mathematics and mathematical tasks" (p. 45). Our view is that beliefs constitute the individual's *subjective knowledge* about self, mathematics, problem solving, and the topics dealt with in problem statements. As Hart (this volume, Chapter 3) points out, the distinction between beliefs and attitudes is murky at best. In an attempt to reduce this murkiness, we have chosen to consider beliefs about self as attitudes, and, thus, to restrict the category we label *beliefs* to beliefs about objects and ideas that are external to the individual. The distinction between beliefs and *objective knowledge* is similarly unclear. The difference between the two rests with the notion that an individual's beliefs may or may not be logically true and may or may not be externally justifiable, whereas knowledge must have both characteristics in addition to being believed by the individual (cf. Kitcher, 1984).

Beliefs often interact with and, at times, shape attitudes and emotions, and beliefs influence the decisions made during problem solving. In our current research, we are especially interested in students' beliefs about the nature of mathematics, in general, and of problem solving, in particular.

## Control

We use the term *control* in much the same way as Schoenfeld (1985). Control refers to the marshalling and subsequent allocation of available resources to solve a problem; more specifically, it deals with executive decisions about planning, evaluating, monitoring, and regulating. Two aspects of control processes are of particular interest to us: knowledge about cognition and regulation of cognition. These two aspects of control make up the psychological construct referred to as *metacognition*. We will not discuss metacognition here, because quite a lot has been written about it in recent years. The reader who wishes to learn more about metacognition and the role it plays in mathematical activity should refer to Garofalo and Lester (1985) or Schoenfeld (1985, 1987).

## Contextual Factors

In recent years, the point has been raised by various cognitive psychologists that human intellectual behavior must be studied in the contexts in which it takes place (Neisser, 1976; Norman, 1981). Because human beings are immersed in a reality that both affects and is affected by human behavior, it is particularly essential to consider the ways in which sociocultural factors influence cognition. In the case of learning mathematics, the development, understanding, and use of mathematical ideas and techniques grow out of social and cultural situations. D'Ambrosio (1985) argues that children bring to school their own mathematics which has developed within their own sociocultural environment. This mathematics, which he calls *ethnomathematics*, provides the individual with a wealth of intuitions and informal procedures for dealing with mathematical phenomena (cf. Carraher, Carraher, & Schliemann, 1985; Rogoff & Lave, 1984). Furthermore, one need not look outside the school or classroom for evidence that social and cultural conditions play an important role in what is learned. It is clear that the sorts of interactions students have among themselves and with their teachers, as well as the beliefs, values, and expectations that are nurtured in school contexts, shape not only what mathematics is learned, but also how it is learned (cf. Cobb, 1986).

The point, then, is that the wealth of sociocultural contexts that make up an individual's reality plays a prominent role in determining the individual's potential for success in solving a given problem. Indeed, it is our view that these factors have a direct influence on each of the other domains.

It is clear that the four types of factors we have just discussed overlap and interact in numerous ways; for example, contextual factors directly influence the formation of attitudes and beliefs, as well as the extent to which an individual is willing or able to engage in regulatory behavior during problem solving. Also, affects and attitudes both influence and are influenced by beliefs. Research on problem solving in classroom settings must give heed to the variety and complexity of noncognitive and metacognitive factors.

## Observations from Our Research

In 1981 we initiated an exploratory investigation of young children's metacognitive awareness as it relates to mathematical performance. This investigation ultimately led to the development of a framework that specifies key points at which metacognitive actions are likely to affect cognition (Garofalo & Lester, 1985; Lester & Garofalo, 1982). The framework has been used to assist us in selecting mathematics tasks, designing instructional activities, studying videotapes of problem-solving sessions, and analyzing students' written work. Since 1981, we have studied the metacognitive behavior of children ranging in age from 6 to 13 years and of university students (Kroll, 1988).

Originally, our present research with seventh graders had two objectives: (a) to assess the metacognitive behaviors of seventh graders and to investigate the role of metacognition in their mathematical problem-solving activity; and (b) to explore the extent to which it is possible to teach these students to be more self-aware of their problem-solving behaviors and to monitor and evaluate these behaviors. As we began to conduct interviews and plan instruction, it became apparent that it would be very difficult to study students' cognitive and metacognitive behaviors without also considering various noncognitive factors. Consequently, in addition to investigating students' metacognitive behaviors, we also planned our study to provide us with information about the students' attitudes and beliefs. In particular, we collected data about the students' interest levels in various types of problems, their self-confidence in solving these problems, their perceptions of problem difficulty, and their beliefs about the nature of mathematics and mathematical problem solving.

Because it is not the purpose of this chapter to report on the details of our work, we do not provide a complete description here. The reader who is interested in such a description can refer to Lester and Garofalo (1986, 1987) or Garofalo, Kroll, and Lester (1987). It is sufficient to say that our study has involved quite a lot of the following: (a) observations of seventh graders solving problems either alone or in pairs; (b) interviews with these same seventh graders about their problem-solving performance and about their attitudes, interests, and beliefs as they relate to doing mathematics and solving problems; (c) observations of problem-solving instruction conducted with two different seventh-grade classes; and (d) analysis of students' written solutions to problems. The seventh graders were students in two mathematics classes in a public middle school in Bloomington, Indiana. The problems included both routine and nonroutine types. All observations of and interviews with individuals and pairs of students were audiotaped and videotaped. Also, nearly all of the classroom instruction in problem solving was audiotaped and videotaped for later analysis.

To illustrate the dominant influence noncognitive factors can have on problem-solving performance, we have extracted from our data some partial accounts, in the form of scenarios, of problem-solving episodes that took place during individual or pair problem-solving sessions or during small-group problem-solving sessions in the classroom.

## Scenario 1: Self-Confidence and Problem-Solving Performance

One class was given the following problem to solve cooperatively in groups of four.

Carla is the drummer in a band. On Tuesday she received her paycheck for playing with the band for one month. She spent 20% of it that day and 50% of what was left on Wednesday. She then had $50 left. How much did Carla receive in her paycheck?

The teacher[3] noticed that one group was not working on the problem. The following exchange took place:

| Teacher: | EN, you had your hand up. Does your group have a question? |
| EN: | We can't do it. We don't know how. |
| Teacher: | Sure you can! Do all of you feel the same way? |
| | (all 4 students in unison: Yeah.) |
| Teacher: | You say you don't know how. Is there something you don't understand about the problem? |
| EN: | Yes, we can't do percents. |

| Teacher: | Come on now. You're kidding me. You worked harder per-cent problems than this in the fifth grade. |
| JE: | But not this year. We haven't done percent yet this year. |

This scenario illustrates a continuing source of frustration for the teacher. Each of the four students in this group had at least average mathematics ability and each was attentive and cooperative in class. The teacher thought that they had the necessary content knowledge to solve this problem or at least that they knew enough to get started. It appeared, however, that they did not believe they could do it. They had convinced themselves that percent problems were too difficult and, consequently, did not even try. These students demonstrated a lack of self-confidence that at times rendered them almost helpless in solving certain types of problems. Their lack of confidence may have been well founded – that is, they may not have understood the concept of percent very well, in which case we see an example of the way in which what one knows, or doesn't know, can affect the formation of attitudes and beliefs.

## Scenario 2: Interest and Problem-Solving Performance

The students in a class were asked to solve the following problem individually without consulting with their classmates.

A pair of day-old mice escaped from their pen and got loose in the basement. It takes mice 1 month to mature and another month to produce a new pair of mice each month. The new pairs of mice grow and reproduce at the same rate as the original pair and each pair has a male and a female. If no mice die, how many pairs of mice will there be after 8 months?

While walking around the classroom as the students worked on the problem, the teacher had the following exchange with a student, AW.

| AW: | Who cares about solving problems about some dumb mice? |
| Teacher: | You are pretty hard to please. What is it about this problem that you don't like? |
| AW: | It's boring! Why can't these problems be about something interesting once in a while? |
| Teacher: | I'll make a deal with you. If you make a list of things that interest you, I'll try to make up some problems using your ideas. But for now I want you to give this one a try. |
| AW: | Well, OK (unenthusiastically), maybe I will. |

Despite above-grade-level performance on standardized mathematics achievement tests, AW exerted very little effort to solve any of the problems the teacher gave to the class. His test scores suggest that his failure was due to something other than a lack of knowledge. AW actually did quite well when he tried; unfortunately, he often did not try at all. To say that he lacked motivation is perhaps an understatement. It was quite common for AW simply to write down any answer to a problem in order to be finished as quickly as possible. AW's lack of interest clearly hampered his development as a problem solver. Because of his disinterest, he rarely bothered to check his own work, consider the reasonableness of his answers, or otherwise engage in good monitoring activities. In this instance, we see a direct link between an attitude (viz., interest) and control.[4]

## Scenario 3: Beliefs About Mathematics and Problem Solving

During individual problem-solving sessions that took place 2 months prior to the beginning of 12 weeks of problem-solving instruction, the following problem was posed.

Atlas Steel makes 4 different types of steel. From a shipment of 300 tons of raw steel, the factory produced 60 tons of Type I, which sold for $60 a ton; 75 tons of Type II, which sold for $65 a ton; 120 tons of Type III, which sold for $72 a ton; and 45 tons of Type IV, which sold for $95 a ton. Raw steel costs $40 a ton. It costs the factory $2500 to convert every 300 tons of raw steel into the 4 types. How much profit did Atlas Steel make on this shipment?

The general approach of one student (WC) to the solution of this and similar problems was guided by a rather strong belief in a very simple fall-back strategy: When in doubt, try all operations one at a time. On the Atlas Steel problem, for example, WC first added weights, then costs, and then gave up on addition. Next she began to multiply costs by weights. After she added up the resulting products, she did not know what to do, so she began dividing the products by 300. WC monitored her work in two ways. She was careful to check the results of her calculations, and she periodically evaluated the reasonableness of her answer so far, as she decided whether or not to try another operation. WC seemed to believe that problem solving is simply a matter of trying all four operations and then picking the result that seems most sensible. This belief overrode any need to truly understand what the problem was all about. On the other hand, this belief also seemed to encourage WC to persevere in the face of a large number of very tedious calculations.

## Scenario 4: Metacognition and Problem-Solving Performance

Figure 6.1 depicts the work turned in by one student (TP) on an in-class assignment. TP spent about 20 minutes working on the problem by herself.

One really cold winter night, the furnace failed at the "Tropicana Fish Shop." Because of this the temperature dropped in the shop and 20 goldfish died. To replace the fish the shop manager bought as many goldfish as he had left. Then as a precaution he divided all the goldfish equally among himself and 3 employees and each person took his share home. The manager got 25 goldfish. How many goldfish were in the shop before the night the furnace failed?

In the preceding scenario, we pointed out that, among other things, WC's metacognitive behavior (in particular, her monitoring) was strongly influenced

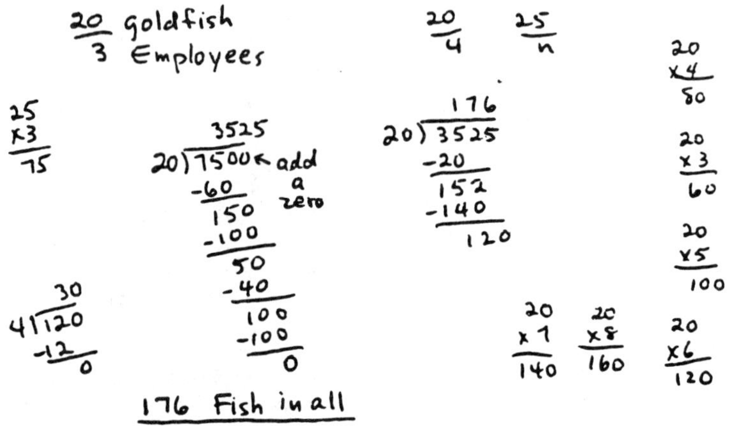

FIGURE 6.1.

by a belief she had about what problem solving involves. TP, even more than WC, evidenced a lack of concern about the reasonableness of her work and of her final answer, but TP's monitoring difficulties seemed to stem more from a lack of knowledge and skills than from a faulty belief.

TP began to work on the problem by recording numbers from the problem at the top of her paper: 20 goldfish, 3 employees. Upon reading further, she apparently understood that there were actually four individuals who took fish home for the night, and thus used the numeral 4 in writing a proportion involving the other numbers in the problem: 20 is to 4 as 25 is to $n$. TP's class had recently completed a unit concerning use of proportions to solve problems, so she apparently was attempting to relate this problem to the technique she had recently been taught. Although such an effort to relate a new problem to an older, more familiar one is laudable, there is no evidence here that TP actually asked, "Do I understand this problem?" or "Is this technique appropriate here?"

TP apparently abandoned her effort to solve the proportion as written. Instead, she multiplied 25 by 3, and proceeded to divide the result by 20. Because 20 would not divide 75 evenly, TP simply added 0s (but no decimal points) to the dividend (neatly recording this move on her paper). There is no evidence here to indicate that TP was at all concerned about the implications of obtaining a non-whole-number quotient for the number of goldfish. (It is quite possible, however, that TP did not even realize that adding zeros meant that the quotient would be a decimal.) TP was, however, apparently concerned about the reasonableness of the answer 3525, because she proceeded to divide this quotient by 20. When she arrived at 176 as the second quotient, she was content with this as a final answer. TP made no effort to check this answer with the original problem conditions.

TP's difficulties with monitoring her work seem to be exacerbated by a lack of basic knowledge and skills. Although she made valiant attempts to identify the knowns and unknowns in the problem and to check her calculations carefully (note the calculations down the right-hand side of the page), her monitoring both of her understanding of the problem and of the reasonableness of her answer were less than adequate.

## Linking Cognitive, Noncognitive, and Metacognitive Factors

Not only do the four types of noncognitive and metacognitive factors influence cognition (i.e., knowledge acquisition and utilization) but, as we mentioned earlier, they also affect each other. In the following paragraphs, we present a few examples of the interrelatedness of some of these factors.

### KNOWLEDGE ABOUT PROBLEM SOLVING CAN AFFECT INTEREST

Prior to the beginning of the 12-week period of instruction, all students were asked to indicate their level of interest in solving a set of five problems that they had just attempted to solve. TT was 1 of only 2 students among 56 who labeled all five problems "very interesting." (Mean score for the five problems was 10.7, responses for each problem were scored as follows: 1 = not interesting; 2 = sort of interesting; 3 = very interesting.) During an interview session, TT said that he liked challenging problems, and because he couldn't "get them at first," he thought these problems were very interesting. After the 12-week instructional period, however, TT's interest in a parallel set of five problems dropped considerably, from 15 to 12; the group mean was now 11.5. Only two students had a greater drop in interest score. A final interview with TT revealed that his loss of interest had resulted from two developments. First, he indicated that the problems were not as challenging for him now because he had several new strategies at his disposal (e.g., guess and check, work backwards, look for patterns). Second, although he now knew how to solve the problems "right away," he thought that "there must be a better way to solve them; maybe with equations or something." It was no longer sufficient for him to figure out how to solve a problem. Now he wanted to solve it in a personally satisfying way. Having developed knowledge of certain problem-solving strategies, he now had begun to look for more elegant methods of solution. This desire for elegance in his solutions influenced what was of interest to him.

### KNOWLEDGE ABOUT PROBLEMS CAN AFFECT SELF-CONFIDENCE

AA was a very good math student who was quite successful in both in-class and homework problem-solving assignments throughout the period of our study with her class; however, AA experienced a 33% decrease in her confidence in her answers to the sets of parallel problems solved before and after the 12 weeks of instruction (her score dropped from a maximum score of 15 to a score of 10; the group means were 12.3 and 12.4, respectively). During a preinstruction interview, AA stated: "Confidence depends on feeling that I understand how to do it, not in getting the right answer." AA also stated: "You're proud of yourself when you get [solve] a weird problem." One possible explanation for her decline is that she became more skeptical of her ability to "get weird problems" as she experienced a broader range of types of problems.

### CONTEXTUAL FACTORS AFFECT BELIEFS ABOUT PROBLEM SOLVING

Lester and Garofalo (1982) found that many third graders believed that all story problems could be solved by applying the operations suggested by the key words present in the story (e.g., *in all* suggests addition, *left* suggests subtraction, *share* suggests division). For some of these students, this belief was very strong. They believed that they had been and would continue to be successful at solving story problems using their key-word strategy. Consequently, they attempted to use such an approach even when it was inappropriate (in fact, even when there were no apparent key words). These third-grade students typically did not bother to attempt to understand relationships expressed in problems, to monitor their actions, or to assess the reasonableness of their answers, because they saw no need to do so. Perhaps the most disturbing thing about this observation is that this belief was a well-founded one. Most of the story problems to which these children had been exposed could be answered correctly by applying their key-word method. Furthermore, some of the children stated that their teachers had taught them to look for key words. Here we have a clear instance of two contextual factors, instructional materials and the teacher, forming a potentially harmful belief.

In our current work with seventh graders, we found another influence at work: grading practices. As a homework assignment, students were asked to write problems that would be interesting to them. Their problems were collected by the teacher, and in a subsequent class period the students were asked to rate each of these problems on a scale from 1 to 5 (very boring to very interesting). After they had completed this task, they were asked to solve any one of the problems (the choice was entirely theirs) that they had rated. The overwhelming majority of the students (about 75%) chose to solve a problem that they had classified as very boring or boring. This was true of both the best and the poorest problem solvers in the two classes. When asked why they had chosen to solve a boring problem, the most common reply was that they wanted to be sure that they "got it right." It appeared that desire for a good grade was stronger than the desire to avoid solving a boring problem.

## Conjectures and Final Comments

At the beginning of this chapter, we suggested that success in solving mathematics problems involves much more than the possession of adequate knowledge. In particular, we posited that various noncognitive and metacognitive factors can have tremendous influence on problem-solving activity, and we presented scenarios and observations taken from our current work to illustrate this claim.

Although at present we can make no confident claims about the specific nature of the influence these factors have on problem solving, we have begun to develop a number of beliefs about the interrelationship among the factors. These beliefs have evolved from our work during the past 7 years with students of vastly different ages and abilities. As a wrap-up to our discussion, we present a few conjectures based on these beliefs.

**Conjecture 1:** An individual's beliefs about self, mathematics, and problem solving play a dominant, often overpowering, role in his or her problem-solving behavior.

There is no question that beliefs affect behavior, particularly behavior associated with learning and doing mathematics; however, beliefs have been all but ignored in studies of problem-solving performance. As a result of our investigations over the past several years, we have become convinced that the beliefs a person holds about his or her ability to do mathematics, about the nature of mathematics, and about problem solving are dominant forces in shaping that person's behavior while engaged in work on a mathematics task. We are particularly interested in the ways in which students' misconceptions influence the formation of attitudes toward mathematics, how and when they monitor their work, and how and when they use the knowledge at their disposal.

**Conjecture 2:** Effective monitoring requires knowing not only *what* and *when* to monitor, but also *how* to do so. Students can be taught what and when to monitor relatively easily, but helping them acquire the skills needed to monitor effectively is more difficult.

Kroll (1988) points out that it is not enough to teach students that they should remind themselves to plan ahead, monitor progress, check their work, and so on; they must also be taught how to do these things. Toward the end of our 12 weeks of metacognition instruction with the seventh-grade classes, it became evident that most of the students were aware that they should monitor their work during problem solving. Furthermore, many of them had become more aware of when to monitor. By and large, however, the students who were very poor monitors of their work initially were still quite weak in this respect at the end of the period of instructional intervention. Although many of these weak students were now aware that they should monitor their progress, most of them had not learned how to do it. We believe that students can be taught to monitor effectively, but progress may be rather slow, especially for those students who have not mastered basic mathematical skills or who have a tendency to be preoccupied with procedural matters to the exclusion of relational and organizational matters (cf. Lesh, 1982).

**Conjecture 3:** Metacognition training is likely to be most effective when it takes place in the context of learning specific mathematical concepts and techniques.

A belief we have held for some time is that problem-solving instruction (or instruction in higher-order thinking in general) should take place in the context of learning a specific body of mathematics content. We hold the same belief about metacognition instruction. In our work with seventh graders, we observed that training in metacognition must be accompanied by attention to instruction in basic mathematics skills and problem-solving strategies. Control processes and aware-ness of cognitive processes develop concurrently with the development of an understanding of mathematical concepts and skills; thus, metacognition training will be most effective when it takes place as a natural part of regular instruction for all mathematics topics. Teachers who expect their students to be metacogni-tive must insure that their students have something to be metacognitive about.

**Conjecture 4:** Persistence is not necessarily a virtue in problem solving.

We have observed that some individuals seem to have a propensity for sticking to a task without regard for getting a correct answer. The effort of WC, whose effort on the "Atlas Steel" problem was discussed earlier, exemplifies such persis-tent, but unreflective, work. Schoenfeld (1985) found such work common among many of the college-age students he studied. He describes students who doggedly pursue a single unprofitable goal as being on a "wild goose chase," and points out that most of his "students were remarkably consistent—and persistent—in pursu-ing wild mathematical geese" (p. 193). Similarly, Kroll (1988), who also studied college-age students, observed a tendency for some students to run in the wrong direction for a long period of time. She noted that these students usually ended up further from home than individuals who had proceeded more slowly and cau-tiously, changing course whenever necessary and frequently reevaluating their orientation. Persistence is a valuable trait, but only if it is accompanied by appropriate monitoring behaviors.

*Acknowledgment.* The research reported in this paper was supported by National Science Foundation Grant No. MDR 85-50346. Any opinions, conclusions, or recommendations expressed are those of the authors and do not necessarily reflect the views of the National Science Foundation.

## Footnotes

[1]This paper is a revised version of a paper prepared for the annual meeting of the Ameri-can Educational Research Association, Washington, DC, April, 1987.

[2]*Knowledge* refers to formal and informal mathematical knowledge, knowledge of heuris-tics (e.g., drawing pictures, looking for patterns, working backwards, devising suitable notation), and knowledge of contextual information present in a problem statement (e.g., knowledge that a quarter is worth 25 cents in a problem involving money).

[3]Frank Lester was the teacher for all the instruction in the seventh-grade classes.

[4]It is tempting to suggest that AW's lack of interest stemmed from a lack of confidence in his mathematical ability; however, his teachers indicated that, although he is intelligent and scores relatively high on standardized tests, he exhibits an apathetic, almost sullen, attitude in almost every classroom situation.

# References

Carraher, T.N., Carraher, D.W., & Schliemann, A.D. (1985). Mathematics in the streets and in the schools. *British Journal of Developmental Psychology, 3*, 21–29.

Cobb, P. (1986). Contexts, goals, beliefs, and learning mathematics. *For the Learning of Mathematics, 6*, 2–9.

D'Ambrosio, U. (1985). *Da realidade a ação: Reflexões sobre educação e matemática* [From reality to action: Reflections about education and mathematics]. São Paulo, Brazil: Summus Editorial.

Garofalo, J., & Lester, F.K. (1985). Metacognition, cognitive monitoring and mathematical performance. *Journal for Research in Mathematics Education, 16*, 163–176.

Garofalo, J., Kroll, D.L., & Lester, F.K. (1987, July). Metacognition and mathematical problem solving: Preliminary research findings. In J.C. Bergeron, N. Herscovics, & C. Kieran (Eds.), *Proceedings of the Eleventh International Conference on the Psychology of Mathematics Education* (Vol. 2, pp. 222–228). Montreal, Canada: University of Montreal.

Kitcher, P. (1984). *The nature of mathematical knowledge*. New York: Oxford University Press.

Kroll, D.L. (1988). *Cooperative mathematical problem solving and metacognition: A case study of three pairs of women*. Unpublished doctoral dissertation, Indiana University, Bloomington.

Lesh, R. (1982). *Metacognition in mathematical problem solving*. Unpublished manuscript (Available from author, WICAT Corp., Orem, UT).

Lester, F.K., & Garofalo, J. (1982, April). *Metacognitive aspects of elementary school students' performance on arithmetic tasks*. Paper presented at the annual meeting of the American Educational Research Association, New York.

Lester, F.K., & Garofalo, J. (1986, April). *An emerging study of sixth graders' metacognition and mathematical performance*. Paper presented at the annual meeting of the American Educational Research Association, San Francisco.

Lester, F.K., & Garofalo, J. (1987, April). *The influence of affects, beliefs, and metacognition on problem solving behavior: Some tentative speculations*. Paper presented at the annual meeting of the American Educational Research Association, Washington, DC.

McLeod, D.B. (1985). Affective issues in research on teaching mathematical problem solving. In E.A. Silver (Ed.), *Teaching and learning mathematical problem solving: Multiple research perspectives* (pp. 267–279). Hillsdale, NJ: Lawrence Erlbaum Associates.

Neisser, U. (1976). *Cognition and reality*. San Francisco: W.H. Freeman.

Norman, D.A. (1981). Twelve issues for cognitive science. In D.A. Norman (Ed.), *Perspectives in cognitive science* (pp. 265–295). Norwood, NJ: Ablex.

Reyes, L.H. (1984). Affective variables and mathematics education. *The Elementary School Journal, 84*, 558–581.

Rogoff, B., & Lave, J. (Eds.). (1984). *Everyday cognition: Its development in social context*. Cambridge, MA: Harvard University Press.

Schoenfeld, A.H. (1985). *Mathematical problem solving*. Orlando, FL: Academic Press.

Schoenfeld, A.H. (1987). What's all the fuss about metacognition? In A.H. Schoenfeld (Ed.), *Cognitive science and mathematics education* (pp. 189–215). Hillsdale, NJ: Lawrence Erlbaum Associates.

Silver, E.A. (1982, January). *Thinking about problem solving: Toward an understanding of metacognitive aspects of mathematical problem solving.* Paper prepared for the Conference on Thinking, Fiji.

Silver, E.A. (1985). Research on teaching mathematical problem solving: Some underrepresented themes and needed directions. In E.A. Silver (Ed.), *Teaching and learning mathematical problem solving: Multiple research perspectives* (pp. 247–266). Hillsdale, NJ: Lawrence Erlbaum Associates.

Webb, N.L., Moses, B.E., & Kerr, D.R. (1977). *Developmental activities related to summative evaluation (1975–1976)* (Tech. Report IV of the Final Report of the Mathematical Problem Solving Project). Bloomington, IN: Indiana University, Mathematics Education Development Center.

# 7
# Information Technologies and Affect in Mathematical Experience

JAMES J. KAPUT[1]

New technologies are causing fundamental changes in the tools and in the contexts in which mathematics is used and applied. The affective consequences of these changes will be as varied as the tools and contexts. No single paper, or book for that matter, can be expected to account for the many changes in the affective dimensions of mathematical experience that will follow from the application of information technology in mathematics learning (Damarin, 1987). This chapter focuses on the "epicenter" of change that relates to the role of information technology as a highly flexible representational medium, one that provides new means of access to mathematical ideas and procedures and new means for representing relationships among ideas and procedures. Other important matters, such as the dynamic and interactive nature of electronic media, are not discussed directly. Additional important social-structure consequences of the use of technology (e.g., changes in group and classroom dynamics, social structure of classrooms, schools, or even formal schooling itself) are also beyond the scope of this chapter. Nonetheless, I hope that by focusing on only a few aspects of change, especially aspects that reflect my immediate experience with certain forms of computer use, I will be able to point to the scope and depth of the revolution that is beginning to take place in mathematics education.

This chapter has four parts, each of which deals with affective issues mainly from the Mandler perspective (this volume, Chapter 1). Sections on the role of multiple, linked representations in changing the affective content of "error" and in building metacognitive control structures are followed by sections about affective issues relating to technology-based strategies and the experience of meaningfulness.

## Changing the Affective Content of Error

The chief characteristic of one new genre of mathematics learning environments is the availability of multiple, linked representations of complex mathematical ideas. The representations are linked in the sense that actions taken in one representation can have immediate and salient consequences in other representations. We are currently investigating the potential impact of such learning environments on the development of skills and key webs of ideas in algebra and in ratio and proportion (Kaput, 1986, in press b; Kaput & Pattison-Gordon, 1987; Kaput, Luke, Poholsky, & Sayer, 1987; Kaput, West, Luke, & Pattison-Gordon, 1988). The overall strategy in applying multiple, linked representations is to take abstract or feature-bare representations that are hard to learn and use, and link them with concrete and feature-rich representations; thus, a student working in the abstract representation can ask for feedback in a more concrete, feature-rich, and "read-able" representation whenever desired. In this section, I first describe a learning environment for skill acquisition in algebra and then a second environment that gives concrete feedback in ratio and proportion problems.

## Skill Acquisition in Algebra

The representation system of formal algebra derives its power from its tight syntax, its rich implicit syntactic structure, and its economy of expression; however, its syntax embodies abrupt shifts of surface structure from one kind of operation to another (e.g., from multiplicative actions to additive actions when exponents are involved). Moreover, algebra requires the use of different styles of substitution rules that determine when one can replace one expression by another, rules that are often distinguished only by subtle and often mixed cues; for example, one type of cancellation applies in multiplicative expressions and another in additive ones (Matz, 1980). Also, there is a shift in substitution rules from operations on fractions as expressions (tacitly representing functions) to operations on fractions in equations. Although the subtlety of each of these shifts is a source of difficulty for students attempting to learn algebra, it also makes possible the economy of expression that is a primary source of the power of algebraic thought (Kaput, in press a).

Students will soon have the option of tracking their actions on formal algebraic expressions through the computer's capacity to present instantly a graph of any single-variable expression as it is written or manipulated (Schwartz, Yerushalmy, & Harvey, 1988). In one window of the computer screen, algebraic expressions (in one variable) are written; in an adjacent window, their graphs appear. The student can act on an expression and, if the expression no longer represents the same function, a new graph will appear on the same axes as its predecessor. For example, suppose a student expands $(x + 3)^2$. If the student makes the common error of leaving out the linear term and writes $x^2 + 9$, then the graph of this new expression appears on the same coordinate plane as the original expression, as depicted in Fig. 7.1a. This is an immediate signal to the student that he or she has changed the expression as a function. The student can then go on to use the system to discover what has been omitted by asking for a plot of the difference between the two expressions (shown in Fig. 7.1b) and to determine the (alpha-numeric) algebraic representation of this difference, which, in this case, is $6x$. Hence, the system does not directly provide information about what was left out in the initial expansion; the student must infer it from other information that he or she generates.

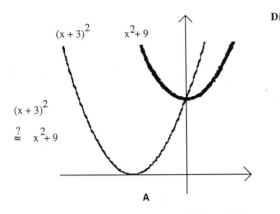
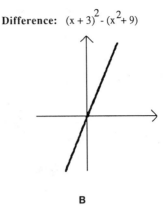

$$\text{Difference:} \quad (x+3)^2 - (x^2+9)$$

FIGURE 7.1.

Another type of task in this learning environment is to change an algebraic expression into a target expression. In this task, the student is given an algebraic expression, its graph, and the graph of the target expression. The student must modify the original algebraic expression in such a way that the graph of the result matches the target graph. Taken together, I suggest that these two types of tasks and the learning context in which they are executed tend to reduce the traditional negative affective content of algebraic error. Traditionally, because the algebraic representation system in referential isolation is so uninformative regarding the appropriateness of actions taken within it, the student needed to be told of the error by some external agent whose authority was grounded outside the structure of the subject matter and whose role was that of a mediating agent between the student and the mathematical representation of the structure of the subject matter. The mediating agent might be a teacher, peer tutor, or a computer tutor (perhaps in an intelligent CAI environment).

The information provided to the student by the mediating agent has three levels. It tells the student *that* an error has occurred, *why* the action was inappropriate, and what an *appropriate* action might be. These three levels of information have sometimes been provided simultaneously in the form of a rule (e.g., a rule for expanding binomials). On the other hand, they have sometimes been provided separately. The first level could have been provided in the form of an explanation based on a numerical evaluation of the two expressions involved: Substitute a number for $x$ in each of the expressions and see that different numbers result (which simply means that the expression was changed as a function because the new one takes on different values from the old one). This particular strategy rests, as recent research is indicating, on the shaky assumption that a student will infer something about an *algebraic* statement from an analogous *arithmetic* one (Lee, 1987). The second level of the explanation (the why level) was usually based on an appeal to the logical structure of the subject matter — in this case, the distributive law; unfortunately, students often do not have a strong grasp of distributivity (Pereira-Mendoza, 1987). The third level amounts to applying this last information to the particular case at hand.

If students understood all of these explanations, they would have approximately the same information as that provided by the graphical feedback in Fig. 7.1. The difference in affective impact between the graphical explanation and the teacher explanation (apart from those differences due to the delay in feedback) is based on the role of the teacher as an agent who mediates the student's interaction with the structure of the subject matter. If some kind of cybernetic tutor were employed, the likely affective impact would simply be based on eliminating the delay and changing the mediating agent from teacher to computer. It has been widely acknowledged that the computer is an affectively neutral agent; however, a computer tutor will yield comparatively positive affective consequences only in those cases in which the teacher had a negative input.

Now suppose that the student did *not* understand the explanation of the error. If the student did not understand any of the three levels of the traditional explanation, then the affective impact is likely to be quite negative. If the feedback provided by the computer learning environment is more accessible than the traditional explanation, then understanding is more likely to occur and, hence, negative affect is less likely to occur. There is reason to believe that the availability of such feedback can significantly increase performance by eliminating actions with unintended consequences; thus, a major consequence of these learning environments is to decrease the likelihood of the situations leading to negative affect.

An entirely analogous graphically oriented environment is being developed for the solution of single-variable equations in which the graphs of each side of the equation are superimposed on the same coordinate graph. The solution of the typical equation is the $x$-coordinate of the point of intersection of the two graphs. Here the user can act on either the algebraic or the graphical representation of an equation and observe the consequences in the counterpart; for example, to add a constant to both sides of an equation yields a vertical displacement of both graphs and no change in the $x$-coordinate of the point of intersection. Multiplying one side of the equation by a constant and not doing the same to the other side, however, is likely to yield a shift in the point of intersection with a horizontal component. Such a shift indicates a change in the solution.

## Concrete Feedback in Ratio and Proportion Problems

Students' difficulties with proportional reasoning have been studied in a variety of contexts. In particular, students' error patterns in solving traditional missing-value problems using the standard algebraic notation have been well documented (e.g., Hart, 1984). We have constructed a four-representation environment for reasoning about ratios (for more detail, see Kaput, in press b; Kaput & Pattison-Gordon, 1987; Kaput et al., 1987). We have constructed a sequence of computer-based learning environments constituting what we have termed a *concrete-to-abstract software ramp* (Kaput & Pattison-Gordon, 1987), in which the concrete end of the ramp engages students in object-based calculations, moving and group-

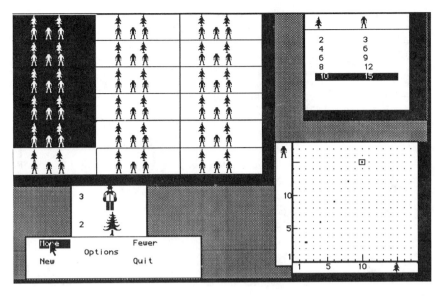

FIGURE 7.2.

ing icons of the type appearing in the upper left window of Fig. 7.2. Other environments, including the one pictured in Fig. 7.2, help establish links among the various representations, three of which appear in Fig. 7.2. (Omitted in Fig. 7.2 is the algebraic equations representation of ratio.) One window each is reserved for iconic, numerical, and coordinate-graph representations of a "per quantity" that the student enters into the computer. The quantity is usually associated with a story. For our purposes, let us assume that we are planting trees in a park so that every two trees will shade three people. After the preliminary student actions that set up the situation, the screen provides an icon window, a table-of-data window labeled with the appropriate icons, and a coordinate-graph window, whose axes are similarly labeled.

In Fig. 7.2, we see the result of five clicks of the *More* button. As the student clicks the *More* button, the cells in the icon window are highlighted, corresponding number pairs are entered in the table of data, and the corresponding points are plotted on the coordinate graph; thus, the intensive quantity is modeled in the coordinate graph as the slope of a line of discrete points. With each click, the latest number pair and the latest point are highlighted to correspond to the number of icons of each type that are highlighted in the icon window. By clicking on the *Fewer* button, the highlighting process proceeds in reverse; however, the previously deposited number pairs in the table and points on the graph remain. Another option enables students to increment or decrement the variables separately. By clicking on the boundary of any of the windows, the student can turn off that particular representation. Turning off representations also helps to control the novelty, when introducing new representations, as well as the amount of

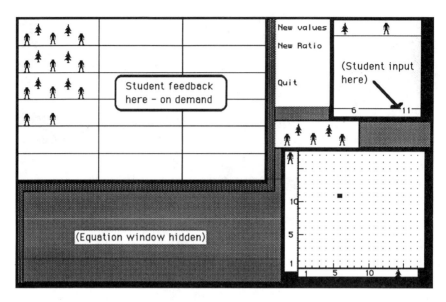

FIGURE 7.3.

information on the screen at any particular time. This arrangement of related windows is referred to as a *linked incrementing environment*.

In another environment designed specifically for solving missing-value problems, the overall appearance and initial actions taken by the student to specify quantities are similar to that of the previous environment, but instead of driving it from the *More* and *Fewer* buttons, the student (a) enters numbers in the table of data, (b) specifies points (by pointing and clicking) in the coordinate graph, or (c) inputs values into an equation. For example, the student provides the corresponding number of an ordered pair when given the other number and the underlying ratio. For a full description of the software environments, see Kaput and Pattison-Gordon (1987) or Kaput (in press b).

By clicking on the boundary to activate the window, the student can view the consequences of his or her input in any window, including the icon window, which is most important for our purposes. There the computer fills in as many cells as possible, duplicating the model cell, so an inappropriate input results in cells that do not match the model, as shown in Fig. 7.3 (there are two "unshaded" people). In this case, the student can try another input. Correct inputs are preserved as pairs in the table of data and points in the coordinate graph to serve as guides for later inputs. As with the graphical feedback in the algebraic environment, the feedback is provided directly by the environment based on the structure of the subject matter, without an intervening or mediating agent. (Of course, no feedback is direct; it is mediated by the student's existing mental structure.) Hence, our earlier analyses apply here as well. Moreover, the concreteness of the representation providing the initial feedback in this case insures that the student's earlier experience with counting and matching actual or screen objects is tapped.

## Ownership and the Contextualization of the Interpretation Process

Another feature of such learning environments that has important affective consequences is the role given to the student in interpreting the consequences of actions. The computer is being used to provide more direct access to additional, presumably more accessible, representations of the subject matter. Ownership of the decision process as well as the repair process (if needed) is in the hands of the student, although a teacher, or even the computer, can be called upon for help if needed. (In our versions of these learning environments, the decision to call for help is a student decision.) The affective consequences of this ownership dimension in such learning environments need to be researched.

A second aspect of mathematical experience in this genre of learning environments concerns the contextualization of actions taken in one representation when their consequences are available for inspection in others. Given that the cybernetically linked representations can be cognitively linked, then the cognitive structures associated with any one representation are enriched by connections to those cognitive structures associated with the other representations. This enrichment amounts to a kind of contextualization that is frequently assumed to occur in the "situated mathematics" of everyday life (Carraher, Carraher, & Schliemann, 1985, 1987; Scribner, 1984). We have produced an extended series of environments that are intended to build these different cognitive structures systematically over a period of several years (Kaput & Pattison-Gordon, 1987). Ongoing classroom-based research is tracking the affective aspects of the resulting experience.

## Building Metacognitive Control Structures

Work with students over the past year suggests that students who have had extensive experience with our various linked-representation environments for reasoning with ratios come to recognize when one representation is appropriate to their needs and when it is not. The knowledge that supports this ability to choose representations for particular purposes seems to be represented in the student's mind tacitly—that is, they do not seem to know that they have it, and they do not know how to describe it explicitly. The tacit nature of this knowledge may result in part from the lack of formal or explicit instruction and the consequent lack of vocabulary for making this knowledge more explicit. It seems fair to say, however, that students do possess tacit knowledge that supports the kinds of problem-solving activities that are commonly described as metacognitive in character; they support the choice of representational contexts in which to do the reasoning.

Although the empirical work involving the use of similar linked-representation environments in algebra is just beginning, it seems reasonable that student ability to move nimbly and appropriately across representations will be enhanced. To be more concrete, we might operationalize this form of knowledge in the following context. Suppose a student has a real-world problem situation that involves setting up a profit function (sales minus costs) as a function of $x$, the number of thousands of items manufactured. Then the process of picking break-even points amounts to finding the roots of this function. A student with a good cognitive control structure that supports access to and evaluation of the available representations might first generate a graph of the function, noting approximately where it crosses the $x$-axis. If there happened to be one such point occurring at an integer $a$ and another occurring at a noninteger, the student might then use the formal algebraic representation to divide the function by $(x-a)$ to obtain a simplified factor of the profit function to analyze it further for recognizable roots. On the other hand, if this did not seem fruitful, the student might call upon a numerical table of data or other numerical methods to find the other root more accurately. The point here is that knowledge of the relative strengths and weaknesses of the various available representations is guiding the problem-solving activity. Whether it is correct to call this metacognitive activity an affective dimension of mathematical problem solving seems a matter of choice.

Another more regulative form of metacognition may be produced by experience with the linked-expression-and-coordinate-graph environment pictured in Fig. 7.1. Research is planned to determine whether work manipulating expressions in this environment, with the immediately available graphical "second opinion," will produce a regulatory awareness of the adequacy of one's algebraic manipulations, especially when one is performing manipulations in the absence of feedback from the coordinate graph. If possible, it will also be interesting to learn how widely such awareness may extend to different forms of symbol-transformation activity not dealt with in that environment.

## Interruptions: Technology-Based Changes in Strategies

McLeod (this volume, Chapter 2) has applied Mandler's theoretical perspective (this volume, Chapter 1) on the relationship between cognitive and affective processes to mathematical problem solving. Mandler's approach is based on the notion that affective experience follows from the interruption of schema-based plans. To the extent that information technologies support strategies that are less "interruptible," they provide means for changing the affective experience of mathematical problem solving. In this section, I discuss some examples that use calculator-based or computer-based strategies that are taken from several beginning-algebra curriculum development projects around the country, beginning with The Ohio State Mathematics Project (Leitzel, 1985; Leitzel & Osborne, 1985). A similar approach is being taken by Usiskin et al. (1985) at the University of Chicago and by Fey, Heid, and Kunkle at the University of

Maryland (Heid & Kunkle, 1988). Others are engaged in substantial curricular revision and development activities along the same lines.

## Word Problems: The Modeling Act

It is widely accepted that two steps are essential for solving a traditional algebra word problem: (a) extract from the text enough information about the quantitative relationships expressed therein to build a cognitive representation of the situation described in the text, and (b) translate those quantitative relationships into the representation system of algebra to solve one or more equations whose solution helps answer the question posed by the original problem. Traditional pedagogical approaches have emphasized the translation process (step b) but have ignored the importance of building a cognitive representation that is sufficiently elaborated to support that translation process; thus, the translation of the problem into algebra is generally a very difficult step. Blockage usually occurs at this step, and is often experienced at what is erroneously assumed to be the very beginning of the solution process. Lacking an appropriate cognitive representation, students are blocked from building first the expressions and then the equations that embody the essential quantitative relationships in the given problem situation. The faulty pedagogy has caused students to generate a plan that is doomed. Using Mandler's (this volume, Chapter 1) theory of emotion, the negative affective consequences of being presented with such an algebra word problem are quite predictable. The schema and plans that result from the reading of the original problem, which necessitate writing an equation and then solving that equation in order to solve the original problem, are immediately blocked, and the student has no choice but to feel frustrated or to engage in trivial translation processes (e.g., key-word matching).

The approach taken by The Ohio State University group (Leitzel, 1985; Leitzel & Osborne, 1985) is to regard such a problem as an occasion for building insight into the original problem situation. The calculator is used to generate concrete data about the problem situation in order to build the requisite cognitive representation. For example, if the problem is to determine the break-even point in sales given that the cost of an item to the seller is $4, then the first step expected of the student is to perform calculations for the cost of two items, three items, four items, and $n$ items. If the price obtained for selling one item is $7, then the student must answer questions such as, "What is the gross profit from two items, three items, four items, and $n$ items?" "Given that net profit is gross profit minus costs and that the fixed costs are $100, what is the net profit after $n$ items?" From here the student can either deduce the solution based on numerical reasoning or, if the problem involves less congenial numbers, set up and solve an appropriate algebraic equation based on the expressions constructed.

The point is that the actions taken are deliberately designed to avoid putting the student in the position of the imminently blockable write-expressions-and-equations-first strategy. Both the schema being built and the goals are different.

The student has something to do when given the problem text that is not nearly as blockable as in the premature-translation strategy. Furthermore, the results of those numerically oriented initial actions, which are intended to build an appropriately rich cognitive representation of the situation, will assist in the translation process. The role of the technology here is to facilitate the initial arithmetic exploration. Thompson (1988) and Schwartz (1988) have constructed modeling environments that are deliberately intended to support such exploration and then to build upon it. The difference between this strategy and the old strategy is analogous to the distinction between walking up a graded trail and scaling a cliff.

## Equation Solving: Reduced Costs of Computation Make General Strategies Available

The histories of the development of many algebraic techniques by the likes of Newton, Napier, Euler, and Laplace reflect the high costs of numerical computation. These mathematicians needed to be able to manipulate equations into particular forms before they could have any hope of "seeing" their structure or solutions. One important characteristic of such techniques is their narrow scope of application; they are very sensitive, for example, to the location of the variable or to the degree of any polynomials involved. To learn algebra, therefore, has meant to become sensitive to all of those often subtle distinctions among forms of expressions and to learn the different techniques that can be applied in each case.

The fourth quarter of the 20th century has seen the floor fall out from under the costs of computation. This reduction in costs is having a major effect on the kinds of strategies that are now "affordable." We can now avail ourselves of very general strategies that are extremely wide in scope and need not distinguish among the particulars of the expressions and equations with which we are dealing. In the matter of solving equations, for example, a common technology-based strategy now being introduced is the guess-and-check strategy. If your proposed solution does not yield equality upon substitution into the equation, then try another number. If the result is farther away from a solution, then try a number in the opposite direction. This strategy is extremely general in its scope of applicability, especially when compared with the analytic strategies that are usually taught. It does not distinguish between solving

$$3.78^x = 41.4 \text{ and } 2.34x^2 + 5.1x = 53.$$

Its only disadvantage is its computational inefficiency, which is tempered by the inexpensiveness of computation. (I include accuracy as an efficiency issue.)

In this chapter, the significant fact is not the reduced cost of computation per se but the fact that it has made possible a strategy that is far less susceptible to blockage because it does not involve techniques with a narrow scope of applications. Given the availability of this kind of strategy, the negative affective consequences of disrupted equation-solving plans can occur much less frequently. Recalling the word-problem context discussed previously, we now have available a less blockable strategy for writing as well as solving equations.

At a higher level of abstraction and formality, symbol manipulators can play the same role, as demonstrated by Fey, Heid, and colleagues (see Heid & Kunkle, 1988). In fact, many contemporary systems now have some version of a *Solve* key. In Heid's work, the symbol manipulator is extended to include a numerical data generator, so that expressions can be compared by having the computer generate adjacent columns of data for each expression. Other tools being developed by Schwartz, Yerushalmy, and Harvey (1988) enable students to produce a graphical representation of the equation and then use graphical-numerical methods to solve the equations (e.g., slide an "*x*-cursor" left or right, and watch an odometer-like reading of the *x*- and *y*- values of the corresponding points on the graph)(Kaput, in press a).

The overall effect of such technological tools is to remove a source of disruption. Such tools do not eliminate the need for some algebraic technique and interpretation of results, nor do they eliminate the possibility of error. Given the kinds of cognitive representation-generating strategies that are possible via appropriate use of computation, however, the other components in the problem-solving act can occur in a much richer and more stable cognitive context, with less likelihood of complete blockage. The long-term affective consequences of problem solving in environments rich in technological tools must be researched, particularly because these tools and contexts will be ubiquitous in the future.

## Affective Issues and the Experience of Meaningfulness

The preceding narrative may leave the reader with the nagging *feeling* that certain dimensions of affective experience are neither encompassed by the schema-plans-blockage perspective, nor accounted for by an attitudes-beliefs-attribution perspective. The dimension that I feel is neglected has to do with the experience of "meaningfulness" of the mathematics with which one is dealing. I believe that this notion of meaningfulness is the phenomenological correlate of the integration and coordination of cognitive representations—representations that fit together "horizontally" and "vertically." To set the stage for this discussion, let us consider it in terms of motivation, especially in school.

Few now deny that school mathematics as experienced by most students is compartmentalized into meaningless pieces that are isolated from one another and from the students' wider world. Symbols are manipulated without regard to the meanings that might be carried, either by referents of the symbols or by actions on them. Theorems are "proved" without the slightest attempt to generate the statements to be proved or to justify the need for proof. This *experienced meaninglessness* of school mathematics devastates the motivation to learn or use mathematics and is entirely incompatible with a view of mathematics as a tool of personal insight and problem solving. This core problem of alienation is compounded by the difficulties inherent in dealing with formal symbols (e.g., algebra, isolated from other knowledge).

The standard historical *curricular* response to student difficulty with maneuvering symbols in isolation is to sequence small pieces of activity that are carefully organized by syntactical features of the symbol system and to isolate this activity from the messiness of applications and wider interpretations. The standard *pedagogical* response is to schedule more empty practice. To deal with the lack of motivation, various trivial sticks and carrots are used, whose triviality is especially well revealed in the kinds of drill-and-practice computer software that found their way into the schools in the early and middle 1980s. The responses made by the students have been highly adaptive; they have used more superficial learning strategies, which result in even more alienation, which in turn feeds the school responses already listed, thus strengthening the feedback cycle.

What are some of the sources of mathematical meaning? From an epistemological perspective, I list four: (a) *transformations within and operations on* a particular representation system; (b) *translations across* mathematical representation systems; (c) most importantly, *translations between* non-mathematically described situations and mathematical representation systems; and (d) the consolidation and reification of actions, procedures, or webs of related concepts into phenomenological objects that can then serve as the bases of new actions, procedures, and concepts at a higher level of organization.

I claim that each of these epistemological sources of mathematical meaning can be described in cognitive structural terms. The first source involves the recognition or elaboration of patterns in one's symbol manipulations — a matching or elaboration of previously learned cognitive structures with new ones generated by the surface forms of the symbols that one is transforming. This is the type of meaning building that predominates in the schools and is *referentially isolated* from other meanings. The second and third sources of meaning are referential extensions of knowledge in the sense that meaning in one representation relates to meaning in another. In other terms, the cognitive structures associated with each representation system become linked, leading to what Lawler has called an *elevation of control* (Lawler, 1981, 1985). The multiple-representation software environments are intended to help support these kinds of meaning building. The fourth source of meaning most directly addresses the missing dimension of meaning noted at the beginning of this section. The "crystallization," or reification, processes by which the fourth source of meaning is achieved are sometimes called *reflective abstraction* in the Piagetian tradition and are central to a constructivist view of mathematics learning (note the deliberate plurals in this sentence). They include the processes by which the act of counting leads to number as a stable conceptual entity, the act of "taking part of" leads to one aspect of rational number, or the process by which the transformation of function-as-procedure for acting on numbers becomes function-as-object that can be added, differentiated, and so forth (Kaput, 1987). Greeno (1983) seems to be describing a similar phenomenon in his discussion of conceptual entities, which, he argues, are stable elements in a representation that are used as the arguments of procedures and can themselves carry attributes or relations.

In the following hypothesis about affective experiences and mathematical meaning, I believe that I am referring to the same positive affect mentioned by McLeod (this volume, Chapter 2) when he refers to von Glasersfeld, Cobb, and Lawler.

**Hypothesis:** Each of the epistemological sources of mathematical meaning amounts to an elaboration, a coherence-improving reorganization, or a transformation of cognitive structures, and each therefore is associated with an affective experience that is usually positive.

Each particular suggestion offered earlier in this chapter that the affective dimension of the given mathematical experience should be researched was offered in the context of this hypothesis. I am suggesting that the building of mathematical meaning has an affective dimension that needs to be better understood in order to determine how or whether that experience might vary in the different categories mentioned. Note that there is not necessarily a direct connection between this issue and information technology, *except* in the sense that the new technologies provide additional means for the successful establishment of any of these sources of mathematical meaningfulness.

*Acknowledgment.* Portions of this chapter report research supported by the Office of Educational Research and Improvement (Contract No. OERI 400–83–0041), but the views expressed are those of the author.

## Footnote

[1]Another paper connected with the project that has given rise to this book (Kaput, in press) deals with attitudes, beliefs, and attributional patterns from the Weiner perspective (Weiner, 1980, 1982) as embodied in survey instruments developed in part by Fennema and colleagues (Fennema, 1982, 1983) and in part by the author. It describes an empirical study intended to determine whether and, if so, to what extent and in what directions, student attitudes toward and beliefs about mathematics are affected by experience with the kinds of technologically based geometry-learning environments provided in classes making intense use of *The Geometric Supposers* (Schwartz & Yerushalmy, 1985).

## References

Carraher, T.N., Carraher, D.W., & Schliemann, A.D. (1985). Mathematics in the street and in schools. *British Journal of Developmental Psychology, 3*, 21–29.

Carraher, T.N., Carraher, D.W., & Schliemann, A.D. (1987). Written and oral mathematics. *Journal for Research in Mathematics Education, 18*, 83–97.

Damarin, S.K. (1987, April). *Dimensions and characteristics of educational software: A framework for research and development.* Paper presented to the annual meeting of the American Educational Research Association, Washington, DC.

Fennema, E. (1982, March). *The development of variables associated with sex differences in mathematics.* Symposium conducted at the annual meeting of the American Educational Research Association, New York.

Fennema, E. (1983, August). *Research on relationship of spatial visualization and confidence to male/female mathematics achievement in Grades 6–8. Final report to the National Science Foundation* (Contract No. SED 78–17330). Madison: University of Wisconsin, College of Education (ERIC Document Reproduction Service No. ED 232 853).

Greeno, J. (1983). Conceptual entities. In A. Stevens & D. Gentner (Eds.), *Mental models.* Hillsdale, NJ: Lawrence Erlbaum Associates.

Hart, K. (1984). *Ratio: Children's strategies and errors.* Windsor, England: NFER-Nelson.

Heid, M.K., & Kunkle, D. (1988). Computer generated tables: Tools for concept development in elementary algebra. In A.F. Coxford & A.P. Shulte (Eds.), *The ideas of algebra, K-12: 1988 yearbook of the National Council of Teachers of Mathematics* (pp. 170–177). Reston, VA: National Council of Teachers of Mathematics.

Kaput, J. (1986). Information technology and mathematics: Opening new representational windows. *Journal of Mathematical Behavior, 5,* 187–207. [Also available as Occasional Paper 86–3. Cambridge, MA: Educational Technology Center, Harvard Graduate School of Education]

Kaput, J. (1987). Toward a theory of symbol use in mathematics. In C. Janvier (Ed.), *Problems of representation in mathematics learning and problem solving* (pp. 159–195). Hillsdale, NJ: Lawrence Erlbaum Associates.

Kaput, J. (in press a). Linking representations in the symbol systems of algebra. In S. Wagner & C. Kieran (Eds.), *Research issues in the learning and teaching of algebra.* Reston, VA: National Council of Teachers of Mathematics. Hillsdale, NJ: Lawrence Erlbaum Associates.

Kaput, J. (in press b). The role of concrete representations of multiplicative structures, Part 2: Ratio and intensive quantity. *Journal of Mathematical Behavior.*

Kaput, J. (in press c). Longer term impacts on student beliefs about and attitudes toward mathematics resulting from *Geometric Supposer* use. In P. Butler, M. Gordon, & J. Kaput (Eds.), *A supposer reader.* Hillsdale, NJ: Lawrence Erlbaum Associates.

Kaput, J., & Pattison-Gordon, L. (1987). *A concrete-to-abstract software ramp for learning multiplication, division, and intensive quantity* (Tech. Rep. No. 87-9). Cambridge, MA: Harvard Graduate School of Education, Educational Technology Center.

Kaput, J., Luke, C., Poholsky, J., & Sayer, A. (1987). Multiple representations and reasoning with discrete continuous quantities in a computer-based environment. In J. Bergeron, N. Herscovics, & C. Kieran (Eds.), *Proceedings of the Eleventh International Conference for the Psychology of Mathematics Education* (Vol. 2, pp. 289–295). Montreal, Canada: University of Montreal.

Kaput, J., West, M.M., Luke, C., & Pattison-Gordon, L. (1988). Concrete representations for ratio reasoning. In M. Behr, C. LaCompagne, & M.M. Wheeler (Eds.), *Proceedings of the Tenth Annual Meeting of the North American Chapter of the International Group for the Psychology of Mathematics Education.* DeKalb, IL: Northern Illinois University.

Lawler, R. (1981). The progressive construction of mind. *Cognitive Science, 5,* 1–30.

Lawler, R. (1985). *Computer experience and cognitive development.* New York: Wiley.

Lee, L. (1987). The status and understanding of generalized algebraic statements by high school students. In J. Bergeron, N. Herscovics, & C. Kieran (Eds.), *Proceedings of the Eleventh International Conference for the Psychology of Mathematics Education,* Vol. 1, pp. 316–323 . Montreal, Canada: University of Montreal.

Leitzel, J. (1985). *Calculators do more than calculate*. Unpublished manuscript, The Ohio State University, Department of Mathematics, Columbus, OH.

Leitzel, J., & Osborne, A. (1985). Mathematical alternatives for the college intending. In C. Hirsch & M. Zweng (Eds.), *The secondary school mathematics curriculum: 1985 yearbook of the National Council of Teachers of Mathematics* (pp. 150–165). Reston, VA: National Council of Teachers of Mathematics.

Matz, M. (1980). Towards a theory of algebraic competence. *Journal of Mathematical Behavior, 3*, 93–166.

Pereira-Mendoza, L. (1987). Error patterns and strategies in algebraic simplification. In J. Bergeron, N. Herscovics, & C. Kieran (Eds.), *Proceedings of the Eleventh International Conference for the Psychology of Mathematics Education* (Vol. 1, pp. 331–337). Montreal, Canada: University of Montreal.

Schwartz, J. (1988). *The algebraic proposer* [Computer software]. Hanover, NH: True BASIC.

Schwartz, J., & Yerushalmy, M. (1985). *The geometric supposers* [A series of four software packages]. Pleasantville, NY: Sunburst Communications.

Schwartz, J., Yerushalmy, M., & Harvey, W. (1988). *The algebra studio* [A series of algebra computer software packages]. Pleasantville, NY: Sunburst Communications.

Scribner, S. (1984). Studying working intelligence. In B. Rogott & J. Lave (Eds.), *Everyday cognition: Its development in social context* (pp. 9–40). Cambridge, MA: Harvard University Press.

Thompson, P. (1988) *Word problem assistant* [Computer software]. Normal, IL: P. Thompson, Illinois State University, Department of Mathematical Sciences.

Usiskin, Z., Flanders, J., Hynes, C., Polonsky, L., Porter, S., & Viktora, S. (1985). *Transition mathematics* (Vols. 1,2). Chicago, IL: University of Chicago School Mathematics Project, Department of Education.

Weiner, B. (1980). The order of affect in rational (attributional) approaches to human motivation. *Educational Researcher, 19*, 4–11.

Weiner, B. (1982). A theory of motivation for some classroom experiences. *Journal of Educational Psychology, 71*, 3–25.

# 8
# Searching for Affect in the Solution of Story Problems in Mathematics[1]

LARRY SOWDER

From the perspective of mathematics education, story problems represent a rather low-level, but important, type of problem solving. Unfortunately, there is little doubt that many people have strong negative feelings about story problems. Anecdotal evidence abounds ("I like math, except for story problems"), and is strengthened by an occasional study. Dutton and Blum (1968), for example, found that the statement "Word problems are frustrating" was one of the more commonly agreed-with statements among 346 students in Grades 6 through 8. It is sobering that even preservice teachers may share such negative feelings; 46% of Smith's (1964) 123 preservice elementary teachers chose the statement "I am afraid of doing word problems" as representative of their feelings.

When and why these negative feelings arise is not clear. For example, British upper elementary students "liked" best a story problem dealing with money, when they ranked a set of exercises that included computation and geometry tasks as well as story problems (Kyles & Sumner, cited in Bell, Costello, & Kuchemann, 1983, p. 244). Are algebra story problems, or multistep problems, or problems with fractions or decimals or percents the culprits? Looking for affect while students solve word problems would appear to be one of the first steps in determining an "etiology" for negative affect toward story problems.

## Students' Strategies for Solving Story Problems

Recent interview studies (Greer, 1987a; Lester & Garofalo, 1982; Noddings, 1985; Sherrill, 1983; Sowder, 1986a; Sowder, Threadgill-Sowder, Moyer, & Moyer, 1983) have suggested strategies that students in the middle grades appear to use in solving story problems. The following catalogue (Sowder, 1986a) is no doubt incomplete, but it does represent the variety of approaches that students use. (Note that S = student, and I = interviewer.)

1. Find the numbers and add.
2. Guess at the operation to be used.

3. Look at the numbers; they will "tell" you which operation to use.
   Example: (Grade 7 student)
   S: Yeah, it looks like a division. . . . It's the numbers, I guess.
   I: Oh, the numbers, huh?
   S: 3 times, 3 goes into 78, that's what it mostly is. Cause if it's like, 78 and maybe 54, then I'd probably either add or multiply. But 3, it looks like a division. Because of the size of the numbers.
4. Try all the operations and choose the most reasonable answer.
   Example 1: (Grade 6 student about a problem involving multiplication)
   I: (After the student discussed rejecting addition for the problem) Did you even think about adding, a couple of weeks ago (during a group test)?
   S: Yeah. I go through every one to see if it would work.
   I: Oh, you do?
   S: And I went through adding, and I saw that that wasn't a good choice, and then I went into subtraction. . .
   Example 2: (Grade 6 student)
   I: . . . what made you think you should divide?
   S: Well, the addition, subtracting, and multiplying didn't look right.
5. Look for isolated "key" words to tell which operation to use (e.g., "all together" would mean add, "left" would mean subtract, "of" would mean multiply).
6. Decide whether the answer should be larger or smaller than the given numbers. If larger, try both addition and multiplication and choose the more reasonable answer. If smaller, try both subtraction and division and choose the more reasonable.
   Example 1: (Grade 7 student)
   S: . . . that (problem) was adding, multiplying. I didn't even bother with them, subtracting or dividing.
   Example 2: (Grade 6 student)
   S: Well, I would think that you have to subtract, because, er, it'd either be a subtract or, um, division, and then the one that sounds right would be the subtraction.
7. Choose the operation whose meaning fits the story.
   Example 1: (Grade 6 student)
   S: There's 24 in each row, and there's 12 rows. And you want to know how many there were, so you'd times.
   Example 2: (Grade 7 student)
   S: . . . I'd probably multiply it . . . I would multiply it, but you could also add it. But it would take you, you would have to add 2.46 three times (adapted from pp. 471–472).

The purpose of this chapter is not to discuss the persistence or the limitations of these strategies. Two points should be noted, however: (a) students who use Strategies 3 through 6 will be successful on many, if not most, one-step story problems in the whole-number curriculum, and (b) few students, even capable ones, give evidence of using the "mature" Strategy 7.

## The Strategies as Responses to Interruptions

The possible origins of the "immature" strategies (1 through 6) might be explained by Mandler's discrepancy theory for emotion (this volume, Chapter 1). Briefly, Mandler describes an emotion as being the combination of an evaluation, perhaps subconscious, that something is not going as desired or as expected (a discrepancy, or interruption) with a consequent response from the autonomic nervous system. If, when attempting to solve a story problem, a student's expectation is to proceed routinely to a solution, as in much of algorithmic mathematics, the situation is ripe for an interruption. Because schooling is oriented toward success, such an interruption ("I don't know exactly what to do, so I may miss this one") might indeed result in some internal response and, thus, produce an emotional response. If subsequent work does result in an incorrect answer, one could expect an even stronger emotional reaction on later story problems. If the lack of success predominates, negative attitudes such as those described earlier are likely to be the result. It is reasonable to assume that students adopt their problem-solving strategies to maximize success – that is, to minimize interruptions.

## Looking for Affect During the Interviews

My collaborators and I have interviewed over 70 students in Grades 4 through 8 while they have been solving story problems (Sowder, 1986a; Sowder et al., 1983). Affect was a special concern in the most recent interviews, which were conducted with eighth-grade students.

Emotional responses may not, of course, give such obvious signs as sweating, gasps, and body tension. Although most of the students interviewed seemed open and forthright, there is also the possibility that "schoolmanship" involves bluff or pretense; hence, overt evidence of emotion during an interview was considered significant and an indication of a relatively strong emotion. Unfortunately, there were only a few such instances in the interviews. The strategies, particularly those that allow false starts, may be so routinely and successfully applied that there is little occasion for strong affect.

### Strategies 1 and 2

Users of Strategy 1, the "automatic adders," often did not give obvious signs of emotional responses during the interviews, except for a lack of enthusiasm for the interview tasks. Indeed, they often tended to be noncommunicative, perhaps in the hope of finishing the interview as soon as possible. Their expectation of success on story problems was probably quite low, so that the tasks no longer carried much potential for emotion. Avoidance behavior during the interview also minimized the emotional content. Only one seventh grader (Strategy 2) was obviously nervous during the interview and acknowledged, with some distress in his voice, that he didn't "know how to decide what to do" in solving story problems. Because the interviews with students who used Strategies 1 and 2 were not very revealing and probably not positive experiences for the students, I stopped interviewing students whose written work suggested that they were using these strategies.

## Strategy 3

The only interviewee who stated, "the numbers tell you what to do" did not show any overt emotional reaction during the interview. This seventh grader was quite voluble and open. Her approach allowed the use of other strategies if the initial answer did not "look" right. If indeed her earlier experience had given her the expectation that she might have to try more than one method before getting a satisfactory answer, then story problems would not present discrepancies for her. Because these immature strategies can be successful, their use may cause minimal violations of expectations of eventual success.

## Strategies 4 Through 6

Users of these strategies often succeed at whole-number story problems. The most striking case of a student who used all four of the operations in solving story problems (Strategy 4) was a seventh-grade girl in a gifted program! Her computational facility and some number sense enabled her to solve the kinds of story problems that she had encountered to date in school.

The key-word strategy (5) is of interest because many teachers teach it without understanding its ultimate limitations. Until recent years, this strategy would ensure success on many of the story problems in textbooks (Pereira-Mendoza [1979] reported a success rate of 97% for the problems in 1 year of a particular text series). I would expect few discrepancies and, hence, little affect, among users of the strategy.

The commonly used Strategy 6 does have good features: The student shows some awareness of reasonable answers and possesses some number sense. Because the strategy is quite effective with one-step problems with whole numbers, users' expectations of success are largely uninterrupted—that is, until fractions and decimals less than one become part of the story-problem curriculum, and "multiplication makes bigger, division makes smaller" (MMBDMS) is no longer a safe guiding principle.

## How Do You Feel About Story Problems?

Because affect was a special consideration during the recent interviews, some students were asked a direct question about how they felt when working on story problems. If a student expressed strong feelings, questions were then asked that would identify early emotional experiences with story problems; however, answers were not especially informative (e.g., "I just say 'okay.' I don't care." "Just regular feelings.") One student acknowledged, "Word problems are fairly confusing." A student who offered, "See, a lot of times when I do word problems, I just think I know the answer, so I'll quickly go over them," responded to the interviewer's query about whether the knowledge that she was going quickly made her nervous: "It doesn't make me nervous." Perhaps she responded this way because she had a fair degree of confidence in her success. There were also no spontaneous attributions (Weiner, 1974); the comment closest to an attribution was a male student's reference to having been in an advanced mathematics group in an earlier grade.

## In situ Interruptions

Many of the interviews have been conducted with seventh and eighth graders who use Strategy 6, and have presented the students with on-the-spot interruptions, when MMBDMS does not work. For example, students almost always succeed on this problem: Cheese costs $2.46 a pound. How much will 3 lb of the cheese cost? When the next problem asks for the cost of 0.82 lb of the cheese, students who use Strategy 6 will note that 0.82 lb is less than a pound, so the answer should be less than $2.46; hence, they opt to subtract or divide. This phenomenon has been studied extensively (e.g., Bell, Fischbein, & Greer, 1984; Ekenstam & Greger, 1983; Fischbein, Deri, Nello, & Marino, 1985; Greer, 1987b; Greer & Mangan, 1986; Mangan, 1986; Sowder, 1986b; Tirosh, Graeber, & Glover, 1986) and has been labeled *nonconservation of operation* by Greer (1987a).

In interviews, we observed that *if* nonconservers were uncomfortable with the answer, they often tried the other operation of the two being considered. If that answer was also unsatisfactory to the student, a common response was confusion. Even when further work showed that multiplication gave the most reasonable answer, some students continued to endorse subtraction or division!

It would be interesting to know whether affect interfered with cognitive adjustment or whether repudiation of MMBDMS caused so great a discrepancy with a long-trusted schema as to be inconceivable (or whether some other reason explains this persistence). Strategy 6 seems to continue even in the face of instruction to combat it. In a small-scale teaching experiment I have carried out with sixth graders, post-instruction interviews made clear that Strategy 6 was still dominant in some students.

## Another Tack: A Semantic Differential

Is *any* identifiable emotion expressed during the solution of story problems? I suspect so, but perhaps story problems are such commonly encountered tasks that any overt expression of emotion during an interview would more likely result from the novelty of an interview situation with a stranger or the request to think out loud. Might another way to detect affect lie in the use of a semantic differential?

Osgood (1976) and Osgood, Suci, and Tannenbaum (1957) hypothesized the existence of a multidimensional semantic space. Using factor analyses on responses to bipolar scales (e.g., clean/dirty), Osgood and his colleagues typically have found "three dominant affective factors (or features): Evaluation (Good/Bad), Potency (Strong/Weak) and Activity (Active/Passive)" (Osgood, 1976, p. 8). Obviously, Hart's warning (this volume, Chapter 3) about the ambiguity of terms comes to mind! The language, if nothing else, led to my use of some of the Osgood scales from the Evaluation factor in an attempt to identify a differential response to story problems—in particular, pairs of problems that invite nonconservation of operation (e.g., Fig. 8.1). It was hypothesized that conservers would have similar responses to the two tasks but that nonconservers would experience difficulty, and thus a discrepancy, on the second problem. This discrepancy would be reflected in different responses on the scales.

2. A woman bought 3 pounds of cheese at $2.46 a pound. How much did she pay for the cheese?

| | | | | | | | | |
|---|---|---|---|---|---|---|---|---|
| ugly | | | | | | | | beautiful |
| clean | | | | | | | | dirty |
| unfair | | | | | | | | fair |
| good | | | | | | | | bad |
| cruel | | | | | | | | kind |
| nice | | | | | | | | awful |
| unpleasant | | | | | | | | pleasant |
| valuable | | | | | | | | worthless |
| bitter | | | | | | | | sweet |

3. Another woman bought 0.82 pounds of cheese at $2.46 a pound. How much did she pay for the cheese?

| | | | | | | | | |
|---|---|---|---|---|---|---|---|---|
| ugly | | | | | | | | beautiful |
| clean | | | | | | | | dirty |
| unfair | | | | | | | | fair |
| good | | | | | | | | bad |
| cruel | | | | | | | | kind |
| nice | | | | | | | | awful |
| unpleasant | | | | | | | | pleasant |
| valuable | | | | | | | | worthless |
| bitter | | | | | | | | sweet |

FIGURE 8.1. Two story problems with semantic differential scales. (The numerals indicate the order on the test.)

TABLE 8.1. Mean responses by group for the paired items (maximum response = 7)

| | Problem | | | Problem | | | Problem | | |
|---|---|---|---|---|---|---|---|---|---|
| | 2 | 3 | t | 5 | 6 | t | 8 | 9 | t |
| Conservers | 5.5 | 4.8 | 5.1* | 5.2 | 4.9 | 2.4** | 5.5 | 4.6 | 4.6* |
| (n = 27) | | | | | | | | | |
| Nonconservers | 6.0 | 4.5 | 6.2* | 5.4 | 4.4 | 5.4* | 5.7 | 4.1 | 5.4* |
| (n = 20) | | | | | | | | | |
| Comparison of | | | | | | | | | |
| between group | | | 2.8* | | | 3.3* | | | 1.9*** |
| mean differences | | | | | | | | | |

*Difference statistically significant at 0.01 level.
**Difference statistically significant at 0.05 level.
***Probability of observed $t$ = 0.067.

Forty-seven preservice elementary teachers enrolled in mathematics courses were given three pairs of items like those in Fig. 8.1 that were intermingled with three fillers. This was followed by a choose-the-operation test to identify possible nonconservers of operation. Of the 47 students, 20 gave evidence of being nonconservers (this perhaps shocking portion is consistent with that found in other studies of adults, such as Mangan, 1986). The results, which contrasted the responses of "conservers" and "nonconservers" to the Osgood scales, are given in Table 8.1. The "positive" extreme of each scale was assigned the numeral 7; the "negative" end, the numeral 1. The first problem in each pair (Problems 2, 5, and 8) involved a multiplier or divisor greater than one; the other problem in each pair (Problems 3, 6, and 9) was matched for context but had a multiplier or divisor less than one.

The nonconservers responded differently to the two items within each pair, as predicted, but so did the conservers, although not so strongly. It is also curious that the nonconservers gave a more positive response to the easier items (Problems 2, 5, and 8) than did the conservers, even though the difference in the means approached significance only for Problem 2 (observed $t$ = 1.99, $p$ = 0.052).

The data suggest that students do have different affective responses to at least the items in such pairs, insofar as the Osgood scales do reflect affect. The differences in the conservers' means for the paired problems are puzzling, because presumably conservers recognize that the problems in a pair are structurally the same. Perhaps even conservers are distracted by the multiplier or divisor less than 1 and experience momentary discrepancies that result in the lower ratings.

## Discussion

In general, the interviewees did not reveal strong emotion during the solution of story problems. The relative "coolness" of any emotion associated with such well-practiced tasks as solving story problems or the overriding affect of an interview situation might explain the apparently nearly emotionless performance of the students. The failure to detect strong emotion in these interviews is also due to the fact that pursuit of affect was a secondary focus in the interviews as well as deficiencies in the interviewer's ability to detect emotion. Nonetheless, in in which nonconservers were confronted with the failure of their usual strategy, there was a clear interruption but also a reluctance to abandon the strategy, even in the face of the disequilibrium. It would be interesting to examine how students work though this discrepancy in the long term.

Despite the lack of overt emotion during the interviews, the use of a semantic differential suggests that there are differential responses within pairs of problems that give the opportunity to exhibit nonconservation of operation. The question of whether affect is primarily responsible for the difference in response is perhaps moot.

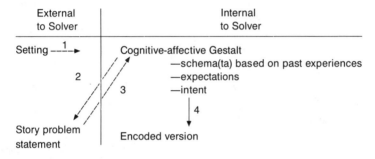

FIGURE 8.2. Some elements involved in approaching a story problem.

The larger-picture analyses by Eccles (1983) and Fennema and Peterson (1983) of possible influences on, and elements in, affect suggest that many factors could play significant roles in students' affect toward mathematics. For example, a student's *intent* in dealing with a task no doubt influences what the student obtains from the encounter (Fig. 8.2). It sometimes is not clear what a student intends as he or she approaches a task. If the student's intent is only to "get done" (Institute for Research on Teaching, 1982; Peterson, Swing, Stark, & Waas, 1984) or to avoid teacher criticism (as may be the case for users of Strategies 1 and 2), surely his or her approaches to tasks, expectations, and, hence, potential for affect (both positive and negative) will differ from those of a student whose intent is to get correct answers to problems or to get correct answers to problems *and* understand (Dweck, 1986; Nicholls, 1983). It is clear that belief systems can shape one's intent (e.g., mathematics can make sense vs. mathematics is a memory-only task). Recent interest in belief systems (e.g., Silver, 1985) is encouraging, particularly as we gain information that can be used in shaping students' intents.

*Acknowledgment.* Many of the interviews were carried out while under support from the National Science Foundation (Grant Nos. SED 8108134 and MDR 8696130). Any opinions, conclusions, or recommendations are those of the author and do not necessarily reflect the views of the National Science Foundation.

## Footnote

[1]This chapter is an expanded version of a paper given at the annual meeting of the American Educational Research Association in April, 1987.

## References

Bell, A., Costello, J., & Kuchemann, D. (1983). *A review of research in mathematical education* (part A). Windsor, England: NFER-Nelson.

Bell, A., Fischbein, E., & Greer, B. (1984). Choice of operation in verbal arithmetic problems: The effects of number size, problem structure, and context. *Educational Studies in Mathematics, 15,* 129–147.

Dutton, W., & Blum, M. (1968). The measurement of attitudes toward arithmetic with a Likert-type test. *Elementary School Journal, 65,* 259–264.

Dweck, C. (1986). Motivational processes affecting learning. *American Psychologist, 41,* 1040–1048.

Eccles, J. (1983). Expectancies, values, and academic behaviors. In J.T. Spence (Ed.), *Achievement and achievement motivation* (pp. 75 –146). San Francisco: W.H. Freeman.

Ekenstam, A., & Greger, K. (1983). Some aspects of children's ability to solve mathematical problems. *Educational Studies in Mathematics, 14,* 369–384.

Fennema, E., & Peterson, P. (1983, April). Autonomous learning behavior: A possible explanation. Paper presented at the annual meeting of the American Educational Research Association, Montreal, Canada.

Fischbein, E., Deri, M., Nello, M., & Marino, M. (1985). The role of implicit models in solving verbal problems in multiplication and division. *Journal for Research in Mathematics Education, 16,* 3–17.

Greer, B. (1987a). Nonconservation of multiplication and division involving decimals. *Journal for Research in Mathematics Education, 18,* 37–45.

Greer, B. (1987b). Understanding of arithmetical operations as models of situations. In J. Sloboda & D. Rogers (Eds.), *Cognitive processes in mathematics* (pp. 60–80). New York: Oxford University Press.

Greer, B., & Mangan, C. (1984). Understanding multiplication and division. In J. Moser (Ed.), *Proceedings of the Sixth Annual Meeting of the North American Chapter of the International Group for the Psychology of Mathematics Education* (pp. 27–32). Madison, WI: University of Wisconsin.

Greer, B., & Mangan, C. (1986). Choice of operations: From 10-year-olds to student teachers. *Proceedings of the Tenth International Conference for the Psychology of Mathematics Education* (pp. 25–30). London: University of London Institute of Education.

Institute for Research on Teaching. (1982, Summer). Do students learn from seatwork? *Communication Quarterly, 5*(1), 2.

Lester, F., & Garofalo, J. (1982, March). Metacognitive aspects of elementary school students' performance on arithmetic tasks. Paper presented at the annual meeting of the American Educational Research Association, New York.

Mangan, M.C. (1986). Choice of operation in multiplication and division word problems. Unpublished doctoral dissertation, Queen's University, Belfast, Northern Ireland.

Nicholls, J. (1983). Conceptions of ability and achievement motivation: A theory and its implications for education. In S. Paris, G. Olson, & H. Stevenson (Eds.), *Learning and motivation in the classroom* (pp. 211–237). Hillsdale, NJ: Lawrence Erlbaum Associates.

Noddings, N. (1985). Small groups as a setting for research on mathematical problem solving. In E.A. Silver (Ed.), *Teaching and learning mathematical problem solving: Multiple research perspectives* (pp. 345–359). Hillsdale, NJ: Lawrence Erlbaum Associates.

Osgood, C. (1976). *Focus on meaning* (Vol. 1). The Hague: Mouton.

Osgood, C., Suci, G., & Tannenbaum, P. (1957). *The measurement of meaning*. Urbana, IL: University of Illinois Press.

Pereira-Mendoza, L. (1979, June). *The mathematics curriculum—the decades ahead.* Paper presented at the Second International Congress on Education, Vancouver, British Columbia.

Peterson, P., Swing, S., Stark, K., & Waas, G. (1984). Students' cognitions and time on task during mathematics instruction. *American Educational Research Journal, 21*, 487–515.

Sherrill, J. (1983). Solving textbook mathematical word problems. *The Alberta Journal of Educational Research, 29*, 140–152.

Silver, E. (1985). Research on teaching mathematical problem solving: Some under-represented themes and needed directions. In E.A. Silver (Ed.), *Teaching and learning mathematical problem solving: Multiple research perspectives* (pp. 247–266). Hillsdale, NJ: Lawrence Erlbaum Associates.

Smith, F. (1964). Prospective teachers' attitudes toward arithmetic. *Arithmetic Teacher, 11*, 474–477.

Sowder, L. (1986a). Strategies children use in solving problems. *Proceedings of the Tenth International Conference for the Psychology of Mathematics Education* (pp. 469–474). London: University of London Institute of Education.

Sowder, L. (1986b). Non-conservation of operation in American algebra students. In G. Lappan & R. Even (Eds.), *Proceedings of the Eighth Annual Meeting of the North American Chapter of the International Group for the Psychology of Mathematics Education* (pp. 90–93). East Lansing, MI: Michigan State University.

Sowder, L., Threadgill-Sowder, J., Moyer, J., & Moyer, M. (1983). *Format variables and learner characteristics in mathematical problem solving* (Final Technical Report). Washington, DC: National Science Foundation. (ERIC Document Reproduction Service No. ED 238 735)

Tirosh, D., Graeber, A., & Glover, R. (1986). Preservice teachers' choice of operation for multiplication and division word problems. *Proceedings of the Tenth International Conference for the Psychology of Mathematics Education* (pp. 57–62). London: University of London Institute of Education.

Weiner, B. (1974). *Achievement motivation and attribution theory.* Morristown, NJ: General Learning Press.

# Part III    Studies of Teaching

# 9
# Young Children's Emotional Acts While Engaged in Mathematical Problem Solving

Paul Cobb, Erna Yackel, and Terry Wood

Several authors in this book have eloquently argued that affective issues in mathematics teaching and learning have long been under-represented themes in research. Our interest in the emotional acts of teachers and children is due in part to Doug McLeod's gentle prodding. In addition, we have recently conducted a teaching experiment in a second-grade classroom for an entire school year. We and others observed that many "nice things" happen in this classroom. The children were generally excited about doing mathematics, were very persistent, did not become frustrated, frequently experienced joy when they completed solutions to personally challenging problems, and did not evidence either embarrassment or jealousy. These observations contrast with the findings of Goodlad's (1983) study of over 1,000 classrooms. He concluded that "affect—either positive or negative—was virtually absent. What we observed could only be described as neutral, or perhaps 'flat'"(p. 467). In fact, the emotional tone of the classroom we observed seemed to contribute substantially to the favorable opinions of the mathematics instruction formed by classroom observers such as parents, other teachers, and administrators. At a minimum, the nurturing of positive emotional experiences for children would seem to have immediate propaganda value.

From the beginning, the classroom teaching experiment gave explicit attention to the children's noncognitive development. The general instructional goals included the promotion of intellectual and moral autonomy (Kamii, 1985) and task-involvement (Nicholls, 1983) as a form of motivation. It was not until we observed and tried to understand what was happening in the classroom that the children's emotional acts in specific situations began to take on greater significance. Initially, we viewed these emotional acts as desirable outcomes and as indicators that things were working out in the classroom as we had hoped. As the year progressed, we have increasingly come to view these acts as essential features of the dynamic, self-organizing social system that characterized life in the classroom.

This chapter is divided into four general sections, the first of which presents the current framework within which we are analyzing the children's emotional acts. We then outline the classroom teaching experiment, focusing primarily on methodology, the rationale for the instructional activities, the classroom organization, and the nature of typical classroom social interactions. Next, we present examples of analyses of the emotional acts of children as they occurred within the context of social life in the classroom. Finally, we consider possible implications of this analysis.

## Emotional Acts, Beliefs, and Social Context

Like other contributors to this book, we subscribe to the cognitive and constructivist approach to emotions. More specifically, we believe that emotional acts are based on cognitive appraisals of particular situations (Mandler, this volume, Chapter 1). From this perspective, emotions are not uncontrollable impulses that just happen to passive sufferers. Instead, "our capacity to experience certain emotions is contingent upon learning to make certain kinds of appraisals and evaluations. . . . It is learning to interpret and appraise matters in terms of norms, standards, principles, and ends or goals judged desirable or undesirable" (Pritchard, 1976, p. 219). The observation that emotions are generated by cognitive appraisals of particular situations makes it possible to talk meaningfully about the construction of emotions.

As Mandler (this volume, Chapter 1) noted, emotional experiences or feelings typically involve the perception of visceral arousal in concatenation with cognitive appraisals. These two aspects of emotions correspond to the distinction between emotion viewed as a state and emotion viewed as an act (Armon-Jones, 1986a). Emotion as state is concerned primarily with the phenomenological aspect of emotional experiences, with emotions as inner feelings. Emotion as act acknowledges the performatory aspect of emotions, which conveys appraisals relating to some standard or value. Our focus in this chapter will be almost exclusively on emotions as acts. As Harré (1986) observed, "emotion words cannot be the names for the [distinct physiological] agitation since it has been clearly demonstrated that qualitatively one and the same agitation can be involved in many different emotions" (p. 8). In other words, introspection does not reveal the existence of a multitude of distinct feelings that correspond to the subtle linguistic differentiation of our vocabulary for discussing emotions (Beford, 1986). The situation is much the same when we focus on the behavior of someone who is having an emotional experience. The behavior associated with anger, for example, differs across people and occasions; conversely, the same observed behavior is interpreted differently depending on the circumstances. The distinction between shame and embarrassment, for example, depends on whether an individual interprets that he or she is at fault in a situation (Beford, 1986). More generally, we agree with Coulter's (1986) claim that we cannot "*identify* the emotion we are dealing with unless we take into account how a person is appraising an object or situation" (p. 121).

### Social Norms

As the example of shame and embarrassment illustrates, emotions have an underlying rationale. Within a culture in general and a local social world, such as an elementary mathematics classroom, in particular, certain emotions are not only warranted in specific situations but, at times, ought to occur (Armon-Jones, 1986a). The teacher in our project capitalized on this aspect of emotions by attempting to teach the children how they ought to feel in particular situations during mathematics instruction. The cognitive basis of emotions is also indicated by the observation that expressions of emotion are open to criticism by reference to the way in which the situation has been interpreted (Armon-Jones, 1986b). We might, for example, attempt to defuse a confrontation by explaining to the angry party that the transgressor did not intentionally infringe on his or her rights. As this example of anger suggests and other analyses show, "the study of emotions . . . will require careful attention to the details of the local systems of rights and obligations, of criteria of value and so on. In short . . . emotions cannot be seriously studied without attention to the local moral order. . . . What is at issue in differentiating emotions are the rights, duties, and obligations of . . . people, *in that culture*"(Harré, 1986, p. 6). In other words, "emotions achieve their qualitative character by being contextualized in the social reality that produces them" (Bruner, 1986, p. 117). It was therefore essential that we pay careful attention to the social norms that the teacher and children mutually constructed when we analyzed emotional acts as they occurred in the project classroom. These acts must be placed in the social context within which they were performed and within which they take on meaning. We will in fact argue that it was because the teacher and children established social norms that contrast sharply with those of typical classrooms that we observed generally desirable emotional acts.

Our concern with the unfolding social world within which we observed emotional acts implies that we will not attempt to abstract particular emotions and treat them as objects that can be studied as independent, detachable objects. We will not, for example, analyze particular emotions, such as joy, but will instead focus on joyful acts as they occur in the concrete world of contexts and activities. Further, because emotional acts have a rationale with respect to the local social order, individuals

can offer an account of their conduct through an examination of "reasons." The causality of internal and external forces becomes irrelevant. Instead of asking, "What caused me to feel ashamed?" the actor asks, "What were my *reasons* for being ashamed?" The scientific observer may be guided by the same perspective (Sarbin, 1986, p. 92).

In other words, emotions are often used to account for someone's actions. Emotion words, therefore, "set the action to be explained, not merely in the context of the rest of the individual's behavior, but in a social context. . . . Emotion words explain by giving one sort of reason for action, i.e., by giving a justification, or partial justification, for it" (Beford, 1986, p. 30). Consequently, we make an inference about the child's appraisal of a situation each time we attribute a particular emotional quality to his or her actions.

The kinds of appraisals that lead to emotional acts follow the occurrence of some perceptual or cognitive discrepancy in which expectations are violated (Mandler, this volume, Chapter 1). McLeod (1985) has noted that this viewpoint is relevant for researchers interested in students' mathematical problem solving because discrepancies or blockages are precisely what characterize true problem solving. Mandler also observed that emotional acts following a discrepancy are less intense if they are considered to be a routine part of life. Similarly, Hundeide (1985) suggested that "what is interesting and problematic is always related to a *standard* of what is taken for granted as typical and normal . . . and *it is the deviations from that standard that create reactions. . . .* As this standard changes, new experiences become habituated and taken for granted"(p. 311). This insight is particularly relevant to our work because all areas of second-grade mathematics, including arithmetical computation, were taught through small-group problem solving that was followed by discussions involving the whole class. Therefore, it is possible that the blockages that occurred during problem solving were not construed as discrepancies by the children because they took the occurrence of blockages for granted — that is, they expected to encounter difficulties. Such expectations would, of course, be expressions of the children's beliefs about the nature of mathematical activity (Cobb, 1986c; Confrey, 1984). The children's beliefs would therefore seem to be a crucial aspect of what Hundeide called the "standard of normative expectancies" (1985, p. 311).

## Sources of Beliefs

If emotional acts are influenced by the nature of the beliefs used to interpret situations, then questions concerning the origin of beliefs immediately become relevant. In our view, students' construction of their beliefs about the nature of mathematical activity involves drawing on paradigm cases to thematize their experiences of doing mathematics. These experiences include interactions with the teacher and peers in the classroom. As Balacheff (1986) noted:

We have to realize that most of the time the pupil does not act as a *theoretician* but as a *practical man*. His job is to give a solution to the problem the teacher has given, a solution that will be acceptable with respect to the classroom situation. In such a context the most important thing is to be effective (p. 12).

A student's realization that he or she has failed to fulfill an obligation can itself give rise to a problematic situation for the student. To the extent that the student wants to fulfill the obligation, situations of this sort can precipitate strong emotions. It should be noted that this case is distinct from that in which the student has experiences that confound what he or she takes for granted. For example, a student might repeatedly fail to meet the teacher's expectations. Failure (as the teacher defines it) is the norm for this student, and he or she expects to fail in the future. Nonetheless, each future failure might be quite traumatic for the student, and give rise to feelings of guilt or embarrassment depending on the circumstances. Here, the discrepancy is with respect to the student's understanding of what he or she is expected to accomplish rather than with respect to his or her understanding of what is typical. In fact, anticipations of typical experiences can lead to emotional acts.

Consideration of students' obligations emphasizes the claim that emotional acts occur within the context of the local social world. This point is crucial to our analysis because the teacher we worked with actively attempted to place the children under certain obligations that differed markedly from those of typical classrooms. In other words, the teacher and children mutually established a nonstandard working consensus (Hargreaves, 1975) or didactical contract (Brousseau, 1984). We are attempting to understand this contract by first identifying the regularities or patterns that occurred in classroom social interactions. For the most part, these patterns are outside the conscious awareness of both the teacher and the students and are repeatedly constructed in the course of interactions (Voigt, 1985). Thus, although the teacher and students did not have a blueprint of the interaction patterns, each subconsciously knew how to act appropriately in specific situations as they arose. By teasing out these patterns, one can infer the largely implicit social norms negotiated by the teacher and students, the norms that structured the local social reality within which they taught and learned mathematics and from which emotional acts derived meaning.

The interaction patterns and associated social norms can themselves be analyzed in terms of both the largely implicit, taken-for-granted obligations that the teacher and students accepted in particular situations and the expectations that they had for each other (Voigt, 1985). Such an analysis simultaneously addresses the teacher's and student's beliefs about their own and each others' roles as they were played out while interacting in the classroom. These beliefs, together with beliefs about the nature of mathematical activity, would seem to constitute a substantial part of the standard of normative expectancies. It is perceived discrepancies with these expectations that give rise to emotional acts.

Thus far, we have argued that emotional acts are generated by cognitive appraisals of situations, and that these appraisals are influenced by the local social order. The appraisals involve a comparison of the interpreted situation with expectations. As Averill (1986) stated, "In cognitive terms, emotions may be conceived of as belief systems or schemas that guide the appraisal of situations, the organization of responses, and the self-monitoring (interpretation) of behavior" (p. 100). With regard to mathematical problem solving, beliefs about the nature of mathematical activity and about one's own and others' roles in the classroom would seem to be particularly relevant. These beliefs are constructed in an attempt to make sense of classroom life during mathematics instruction. Our emphasis on the cognitive basis of emotions as acts in no way denies that people feel emotions or that they may, on occasion, feel gripped by a particular emotion that is beyond their control. In our view, these sometimes intense emotional experiences are generated by subjective cognitive interpretations of particular situations.

As a further point, emotional acts are "functional in that they are constituted and prescribed in such a way as to sustain and endorse cultural systems of beliefs and value" (Armon-Jones, 1986b, p. 57). In other words, emotional acts play a role in the development and regeneration of the obligations and expectations that regulate activity in such situations as a classroom during mathematics instruction. We have noted that emotional acts have a rationale, and that this rationale derives in part from the local order (or at least the acting individual's understanding of it as represented by his or her beliefs). Emotional acts that are warranted in one social context might well be completely inappropriate in another. This would seem to be the case with the second-grade classroom we studied when compared with a more typical classroom. Occasions when the children in our class make socially appropriate emotional acts (e.g., rushing excitedly to the teacher to tell her about their solution to a personally challenging problem) served to sustain and endorse the beliefs about mathematical activity and themselves that the teacher had attempted to nurture. In effect, these emotional acts serve the social function of helping to keep the communal story about doing mathematics alive. The emotional sentiment of their actions indicated not only their sincerity but also the significance and importance they attributed to the story. Further, the social appropriateness of their emotions made them autonomous adherents to the story in a way that mere rational comprehension of it would not (Armon-Jones, 1986b, p. 81). This simultaneously served to regulate socially undesirable behavior. It is one thing for a child to understand that some act would transgress particular norms, and quite another to know that he or she will feel guilty after doing it.

Thus far, we have claimed that appropriate emotional acts sustain social norms. Conversely, socially inappropriate emotional acts indicate either that the student has misinterpreted others' intentions or that the student's beliefs are incompatible with social norms that are acceptable to the teacher and other students. Because these acts are open to criticism by reference to norms, their occurrence constitutes opportunities for the teacher and other students to initiate a dialogue about beliefs and values. Further, socially undesirable acts can be criticized by explaining that their construal by others will constitute a cognitive basis for negative emotions. In other words, the student can be told that his or her actions in this social context will probably make other people feel bad. Finally, the teacher does not have to wait for students to evidence a particular emotion, whether positive or negative, before ascribing that emotion (Armon-Jones, 1986b). For example, delight can be prescribed in those situations in which the teacher believes that students *ought* to feel pleased with their accomplishments (e.g., situations in which they persisted and solved a personally challenging problem through their own efforts). In attempting to understand the prescription and feel delighted, students have the opportunity to reorganize their beliefs about the nature of mathematical activity. In the last analysis, however, it is the students who have to make construals that constitute a cognitive basis for delight.

In summary, emotional acts not only support but can actually play a role in the teacher's and students' mutual construction of social norms. This certainly appears to be the case with the classroom that we observed. We discuss the essential features of the teaching experiment in the following section before analyzing examples of emotional acts as they occurred in the classroom.

# Overview of the Teaching Experiment

The teaching experiment was conducted in one second-grade public school class-room for the entire school year as part of a 3-year research and development project. The experiment had a strong pragmatic emphasis in that we were respon-sible for the mathematics instruction of the 20 children in the classroom; thus, we had to accommodate a variety of institutional constraints while developing and implementing instructional activities in a manner compatible with construc-tivist learning theory (Cobb & von Glasersfeld, 1984; Piaget, 1970, 1980; von Glasersfeld, 1983, 1984). Not surprisingly, the constraints profoundly influenced the ways in which we attempted to develop a form of practice compatible with both constructivism as a general theory of knowledge and specific models of early number learning (Steffe, von Glasersfeld, Richards, & Cobb, 1983; Steffe, Cobb, & von Glasersfeld, 1988). We were fortunate in that the classroom teacher was a member of the project staff. Her practical wisdom and insights proved to be invaluable.

## *Methodology*

The teaching experiment conducted in the classroom is a natural extension of the constructivist teaching experiment methodology, in which the researcher interacts with a single child and attempts to guide the constructive activities of the child (Cobb & Steffe, 1983; Steffe, 1983). In our view, the two meth-odologies are appropriate for different phases of a research program (Cobb, 1986a). Both methodologies allow the researcher to focus on the critical moments when children make cognitive restructurings as they develop increas-ingly powerful ways of knowing mathematics. In the case of the classroom teach-ing experiment, these restructurings occur as the children interact with the teacher and their peers rather than with the researcher. The methodology also allows the researcher to address a variety of related issues; the most important is to embed the children's learning of mathematics in social context. To this end, the mutually constructed social norms are analyzed in terms of the teacher's and children's obligations and expectations in the classroom. The children's emo-tional acts both take meaning from and contribute to the construction and con-tinual regeneration of these norms.

The classroom teaching experiment also bears certain resemblances to a type of Soviet experiment that Menchinskaya (1969) called a macroscheme: "Changes are studied in a pupil's school activity and development as he [or she] makes the transition from one age level to another, from one level of instruction to another" (p. 5). However, there is a crucial difference between our approach and that of the Soviet researchers. Typically, Soviet investigators construct the instructional materials before the experiment begins (e.g., Davydov, 1975). We, in contrast, developed samples of a wide range of possible instructional activities in the year preceding the experiment, but the specific activities used in the classroom were developed, modified, and in some cases, abandoned while the experiment was in progress. To aid this process, two video cameras were used to record every mathematics lesson for the school year. While working in groups, eight children faced the cameras. Consequently, it was possible to record the problem-solving activity of approximately half of each of the four pairs of children. Initial analyses of both the whole class dialogues and the small group interactions focused on the quality of the children's mathematical activity and learning as they completed and discussed their solutions to specific instructional activities. These analyses, together with the classroom teacher's observations, guided the development of instructional activities and, on occasion, changes in classroom organization for subsequent lessons. Thus, the processes of developing materials, conducting a formative assessment, and developing an initial explanation of classroom life were one and the same.

## Rationale for Instructional Activities

From the constructivist perspective, substantive mathematical learning is an active problem-solving activity (Cobb, 1986b; Confrey, 1987; Thompson, 1985; von Glasersfeld, 1983). This is the case even when students receive direct recitation instruction. In this context, substantive learning refers to cognitive restructuring as opposed to accretion or tuning (Rumelhart & Norman, 1981). Consequently, our primary focus as we developed, implemented, and refined instructional activities was on that aspect of cognitive development that is both the most significant and the most difficult to explain and influence.

At the risk of oversimplification, an immediate implication of constructivism is that mathematics, including the so-called "basics," such as arithmetical computation, should be taught through problem solving. Admittedly, the application of an efficient computational algorithm, once constructed, can be a routine task for a child; however, the process of constructing such algorithms is characterized by active problem solving (Cobb & Merkel, in press; Kamii, 1985; Labinowicz, 1985) in which conceptual and procedural developments should ideally go hand in hand (Cobb, Wood, & Yackel, in press a; Hiebert & Lefevre, 1986). The concern with mathematical problem solving does not mean that the instructional activities necessarily emphasize what are traditionally considered to be problems — stereotypical textbook word problems. The general notion that ready-made problems can be given to students is highly questionable; instead, teaching through problem solving acknowledges that problems arise for students as they attempt to achieve *their* goals in the classroom. The approach respects that students are the best judges of what they find problematic and encourages them to construct solutions that are acceptable to them given their current ways of knowing. The situations that children find problematic take a variety of forms, including resolving obstacles or contradictions that arise when they use their current concepts and procedures, accounting for surprising outcomes (particularly when two alternative procedures lead to the same result), verbalizing their mathematical thinking, explaining or justifying solutions, and resolving conflicting points of view. As these examples make clear, genuine mathematical problems can arise from classroom social interactions as well as from attempts to complete the instructional activities. (A more detailed discussion of problematic situations can be found in Cobb, Wood, & Yackel, in press b.)

## Classroom Organization

The instructional activities were of two general types: teacher-directed whole-class activities and small-group activities. To the extent that any lesson can be considered typical (Erickson, 1985), the teacher would first spend at most 5 minutes introducing the small-group activities to the children to clarify the intent of the activities. She would, for example, ask the children what they thought a particular symbol meant or ask them how they interpreted the first activity. In doing so, she did not attempt to steer the children towards an official solution method, but instead tried to ensure that the children's understanding of what they were to do was compatible with the intent of the activity as she understood it. Any suggested interpretation or solution, however immature, was acceptable provided it indicated the child had made appropriate suppositions.

Next, a child gave an activity sheet to each group of two or, occasionally, three children. As the children worked in groups for perhaps 25 minutes, the teacher moved from one group to the next, observing and frequently intervening in their problem-solving efforts. Children moved around the classroom on their own initiative. Some went to a table to get one of the available manipulatives that they had decided was needed. Others got additional activity sheets or perhaps a piece of scrap paper. Some groups completed four or five activity sheets while others completed only one, with the teacher's assistance. Finally, the teacher told the children when there was only 1 minute of work time remaining. Most of the children began to put away the manipulatives and prepared for the discussion of their solutions.

The teacher started the discussion by asking the children to explain how they solved the first activity. Sometimes she asked follow-up questions to clarify the explanation or to help the child reconstruct and verbalize his or her solution. Occasionally, a child would become aware of a problem with his or her solution while explaining it to the class. Because of the accepting classroom atmosphere, the child did not become embarrassed or defensive but might simply say "I disagree with my answer" and sit down. It was immediately apparent that the teacher accepted all answers and solutions in a completely nonevaluative way. If, as frequently happened, children proposed two or more conflicting answers, she would frame this as a problem for the children and ask them how they thought the conflict could be resolved. Children volunteered to justify particular answers and, typically, the class arrived at a consensus. On the rare occasions when they failed to do so, the teacher wrote the activity statement on a chalk board so that the children could think about it during the following few days. Although the discussion might have continued for 15 or 20 minutes, the time was sufficient to consider only a small proportion of the activities completed by some groups (the children have much to say about *their* mathematics). Eventually, the teacher terminated the discussion owing to time constraints. She collected the children's activity sheets and glanced through them before distributing them to parents; however, she did not grade their work or in any way indicate whether or not their answers were correct.

In the remaining 10 minutes of the 1-hour lesson, the teacher introduced a whole-class activity and posed one or more questions to the children. She was again nonevaluative when the children offered their solutions and, as before, attempted to orchestrate a discussion among the children.

## Classroom Social Interactions

The teacher's overall intention as she led the class in discussions was to encourage the children to verbalize their solution attempts. Such dialogues give rise to learning opportunities for children as they attempt to reconstruct their solutions (Levina, 1981), distance themselves from their own activity in an attempt to understand alternative points of view (Sigel, 1981), and resolve conflicts between incompatible solution methods (Perret-Clermont, 1980). However, the teacher's expectation that the children should verbalize how they actually interpreted and attempted to solve the instructional activities ran counter to their prior experiences of class discussions in school (Wood, Cobb, & Yackel, 1988). The teacher, therefore, had to exert her authority in order to help the children reconceptualize their beliefs about both their own roles as students and her role as the teacher during mathematics instruction. She and the children initially negotiated obligations and expectations at the beginning of the school year, which made possible the subsequent smooth functioning of the classroom. Once established, this mutually constructed network of obligations and expectations constrained classroom social interactions in the course of which the children constructed mathematical meanings (Blumer, 1969). The patterns of discourse served not to transmit knowledge (Mehan, 1979; Voigt, 1985) but to provide opportunities for children to articulate and reflect on their own and others' mathematical activities.

The teacher's and students' mutual construction of social as well as mathematical realities was reflected in the dual structure of classroom dialogues. At one level, they talked about mathematics; at another level, they talked about talking about mathematics. As in a traditional classroom, the teacher was very much an authority figure who attempted to realize an agenda. The difference resided in the way she expressed her authority in action (Bishop, 1985). When she and the children talked about talking mathematics, the teacher typically initiated and attempted to control the conversation. When they talked about mathematics, however, she limited her role to that of orchestrating the children's contributions. These two conversations were conducted at distinct logical levels (Bateson, 1973), one in effect serving as a framework for the other; therefore, it makes sense to say that the teacher exerted her authority to enable the children to say what they really thought.

The following dialogue, which is taken from an episode that occurred at the beginning of the school year, illustrates the development of the basic pattern of interaction during whole-class discussions. The dialogue centers on word problems that were shown on an overhead projector.

| | |
|---|---|
| Teacher: | There are 6 more tulips behind the rock [4 are in front of the rock]. How many in all? What do we have to do to figure it out? Kara. |
| Kara: | 10. |
| Teacher: | How did you figure this out? . . . Kara, how did you get your answer? |
| Kara: | I got 6 and then added 4 more. |
| Teacher: | She got 6 and added 4 more. Did anybody else get that answer or maybe did it a different way? Yes, Andrew. |
| Andrew: | 11. |
| Teacher: | You had 11. How did you get 11? |

Because the teacher wanted to make the children aware that she respected their solutions to the problems and, at least implicitly, help them come to believe that mathematical solutions should be justifiable, her response at this point deviated distinctly from typical patterns of classroom interaction (Mehan, 1979). Instead of evaluating Andrew's response, she asked him for an explanation.

| | |
|---|---|
| Andrew: | Well. Um. Wait a minute. There would be 10 flowers. |
| Teacher: | How did you discover that? |
| Andrew: | If it was 4 flowers, 6 flowers in front and 6 flowers in back. That would equal up to 12. If you took 2 away to make 4 in front and 6 in back it would make 10. |
| Teacher: | Did anybody else do it a different way? Lisa. |
| Lisa: | 5, 6, 7, 8, 9, 10 (counting on her fingers). |

The teacher's nonauthoritarian, nonevaluative role as she orchestrated the children's contributions to discussions about mathematics was in contrast to her directive interventions when she initiated and guided conversations in which she and the children talked about talking about mathematics. The following incident occurred later in the same episode.

| | |
|---|---|
| Teacher: | Take a look at this problem. "The clown is first in line. Which animal is fourth?" Peter. |
| Peter: | The tiger. |
| Teacher: | How did you decide the tiger? . . . Would you show us how you got the fourth? |
| Peter: | (Goes to the screen at the front of the room.) I saw the clown and then . . . (He counts the animals.) Oh, the dog [is fourth]. (He hesitates.) Well, I couldn't see from my seat. (He looks down at the floor.) |

Teacher:    Okay. What did you come up with?
Peter:      I didn't see it. (He goes back to his seat quickly.)

The teacher realized that in making Peter obliged to explain his solution, she had put him in the position of having to admit that his answer was wrong in front of the entire class. Peter construed this as a situation that warranted embarrassment. Crucially, the teacher was immediately directive in her comments as she talked about talking about mathematics.

Teacher:    That's okay, Peter. It's all right. Boys and girls, even if your answer is not correct, I am most interested in having you think. That's the important part. We are not always going to get answers right, but we want to try.

She expressed her expectations for the children by telling them how she as an authority interpreted the situation. She emphasized that Peter's attempts to solve the problem were appropriate in every way, and simultaneously expressed to the other children her belief that it was more important in this class to think about mathematics than to get right answers.

As the preceding example illustrates, it was the teacher who typically initiated the mutual construction of obligations and expectations in the classroom. In doing so, she simultaneously had to accept certain obligations for her own actions. If she expected the children to honestly express their current understandings of mathematics, then she was obliged to accept their explanations rather than to evaluate them with respect to an officially sanctioned method of solution; thus, the teacher had obligations to the children, just as they did to her. The interlocking obligations and expectations established by the teacher and her class constituted a trusting relationship. The teacher trusted the children to resolve their mathematical problems, and they trusted her to respect their efforts.

These obligations and expectations influenced the children's activity as they worked in small groups, because the children anticipated that they would have to explain and, if necessary, justify their solutions. In addition, the teacher attempted to place the children under the obligation of solving problems in a cooperative manner and of respecting each others' efforts when they worked in groups. As with the whole-class setting, the teacher was explicit about what she expected of the children as they worked together in small groups (Wood & Yackel, 1988). The children were obligated to explain their solution methods to each other and, at a minimum, to agree on a common answer. When possible, the teacher also encouraged them to agree on a solution method.

Our discussion of obligations and expectations directly addresses the children's evolving beliefs about their own and the teacher's role. In addition, we have implicitly dealt with certain aspects of the children's beliefs about the activity of doing mathematics, another crucial aspect of the standard of normative expectancies. For example, the whole-class obligations nurtured the belief that mathematical activity should be explainable, justifiable, and rationally grounded. Further, the children had the opportunity to view mathematics as an activity under their control rather than as disembodied, objectified, subject matter content. The children also came to realize that mathematical problems can have multiple solutions. With regard to small-group work, the children's acceptance of the obligation to think through their problems for themselves, as indicated by their persistence, evidenced the belief that it sometimes takes hours (literally) rather than minutes to solve mathematical problems.

## Examples from the Classroom

The teacher's insistence that the children reach a consensus as they worked in groups meant that the children had two distinct types of problems to solve. The first concerned the mathematical problems that arose as they attempted to complete the instructional activities; the second involved the negotiation of a viable cooperative relationship that would make it possible for them to solve *their* mathematical problems. Emotional acts occurred as the children attempted to resolve each of these two types of problems. In this chapter, our primary focus is on the emotional acts related to doing mathematics rather than those related to the problem of cooperation. With regard to the pragmatics of the classroom, a basic level of cooperation was necessary if the children were to construct mutually acceptable solutions to the mathematics activities (Cobb, Wood, & Yackel, in press b). Consequently, children were assigned to different partners if they were unable even with teacher intervention to solve the problem of cooperation over an extended period of time. This became increasingly rare as the year progressed.

### Construction of Classroom Norms

The emotional acts displayed in a specific situation depend on the interpretation given to the situation, which in turn depends on the social norms (or at least understanding of the norms, as represented by beliefs). The teacher played a crucial role in initiating the mutual construction of the classroom norms. On numerous occasions, she brought specific situations to the attention of the whole class and asked the children to elaborate on their feelings in that situation. In effect, the teacher told the children how they ought to construe the situation. For example, the following episode occurred at the beginning of a class discussion that followed small-group work. One pair of children volunteered that they had spent the entire 20 minutes allocated to group work on a single problem.

| | |
|---|---|
| Kara and Julie: | Because at first we didn't understand it. |
| Teacher: | How did you feel when you finally got your solution? |
| Kara and Julie: | Good! |

Kara and Julie's excitement at having solved the activity was indicated by the manner in which they stood up and almost jumped up and down on the spot during this interchange with the teacher. Julie went on to explain that she had wanted to go on to another activity but that Kara had insisted that they continue working until they had solved the problem. By calling the attention of the entire class to this incident, the teacher demonstrated that the two girls had construed the situation appropriately. In doing so, she implicitly ruled out as inappropriate construals that would lead to feelings of embarrassment, inadequacy, or stupidity at having completed only one activity, even though a number of groups had completed several. At the same time, she illustrated that the classroom social norms are such that "feeling good"—that is, a feeling of satisfaction and pride in one's own accomplishments—is a socially acceptable emotional response to situations in which the children fulfill their obligation of persisting in problem solving and complete a personally challenging task. By way of contrast, the teacher never drew attention to a group that had completed a relatively large number of activities.

The teacher repeatedly emphasized that figuring out problems for yourself ought to make you "feel good." The following episode occurred less than a week after the one just reported. Children had been working in pairs on several problem-solving tasks and found one activity particularly challenging. The following dialogue is part of the total-class discussion that followed small-group work.

Teacher:    One of the problems you're having is figuring out what
            you're expected to do. (The teacher then said that they
            would talk about the problem to clarify what is expected.)
Andy:       Wow! I figured it out.
Teacher:    What if someone asks you for the answer?
Andy:       I won't tell them.
Teacher:    Good for you. Let them figure it out for themselves and get
            the enjoyment out of figuring it out for themselves. It
            makes us feel so good when we do something.

In both of these episodes, the teacher used the children's emotional acts to endorse and to sustain the construals from which the emotional acts were derived. The children's positive emotional acts indicated that their understandings of the classroom norms were appropriate and, through the teacher's interventions, served to sustain and perpetuate the norms.

The teacher also capitalized on situations that arose naturally in the classroom in an attempt to indicate to the children that some acts were socially undesirable in this classroom because they might make people feel bad. The following episode is extracted from a dialogue that occurred at the beginning of a whole-class discussion about an activity that was designed to encourage the construction of increasingly sophisticated concepts of 10 and increasingly sophisticated thinking strategies. The episode began with the teacher talking about incidents that she had observed and that she considered inappropriate with respect to the social norms of the classroom.

Teacher:    Now another thing I noticed was happening, and it is some-
            thing I *don't* like and I *don't* want to hear because it makes

me feel bad, and if it makes me feel bad it probably makes someone else in here feel bad. It's these two words. (She writes "that's easy" on the chalkboard and draws a circle around the phrase.) These words are no, no's starting today. What are these two words, Mark?

Mark:       That's easy.
Teacher.    That's right. When we are working in math, I've had kids come up to me and say, "Oh, that's easy!" Well, maybe I look at it and say, "Gee whiz, I don't think that's very easy." How do you think that's going to make me feel?
Brenda:     Bad.

The teacher listened as several other children offered their suggestions and then explicitly told them her interpretations.

Teacher:    . . . It hurts my feelings when someone says, "Oh that's easy!" (She points to the words on the board.) When I am struggling and trying so hard, it makes me feel kind of dumb or stupid. Because I am thinking, gosh, if it's so easy why am I having so much trouble with it?

She closed the conversation by explicitly telling the children that saying "that's easy" violates a social norm.

Teacher:    So that's going to be something we are not going to say. You can think it if you like, but I don't want you to say it out loud because that can hurt other people's feelings. And what's one of our rules in here? It's to be considerate of others and their feelings.

## Interpreting Situations

Another way in which the teacher initiated the mutual construction of the social norms that determined the appropriateness of emotional acts was to articulate alternative interpretations of situations. In the following example, the children were working on an activity about time in which they were to answer questions individually about their daily schedules. Andy had completed his own answers and began to tell John what to write for his answers, but John refused to accept Andy's suggestions. At this point, the teacher arrived to observe the pair's activity.

Teacher:    How are you gentlemen doing here? Okay, whose side is whose here? . . . Okay, it says, "What time do you each go to bed?" You [John] go to bed at 8 o'clock. You [Andy] go to bed at 8:30. And what time do you get up? You both get up at 7. How many hours do each of you sleep?
John:       I'm still figuring that out.

| Andy: | I sleep 11 hours. |
|---|---|
| Teacher: | (To Andy.) Will his [John's] time be the same time as your time? |
| Andy: | Un-uh. I told him 11½ hours for him. But he doesn't believe me. |
| John: | I don't know. |
| Teacher: | (To John.) You just don't know. (To Andy.) It's not that he doesn't believe you. Maybe he's just not really sure. |
| John: | I'm not really sure. (The teacher leaves and John writes in his answer.) |

Andy tried to meet his obligation of helping his partner work out the solution to the problem. In Andy's view, John rejected his attempt to be helpful, and Andy showed severe irritation. But the teacher suggested an alternative explanation. In this classroom, children were also obligated to think things through for themselves and try to make sense of them. John was still trying to figure out how to solve the problem and did not want to simply accept Andy's answer. By proposing this alternative explanation, the teacher encouraged Andy to construe the situation in a different way. In effect, she told Andy that John's rejection of his answer did not warrant irritation and that he ought to consider the possible reasons for John's rejection.

The teacher also attempted to defuse confrontations that arose in the total-class setting by suggesting interpretations that would make a child's emotional acts unwarranted. The following brief episode is taken from a whole-class discussion in which one child complained angrily that another had provided the answer to a problem.

| Teacher: | Well, I think what was happening — I think he [Ronnie] was proud he had the answer because we had worked that out. |
|---|---|

In this situation the "offending" child, Ronnie, was the weakest student in the class. The teacher's comment reflects an interpretation that he may have been so proud that his group had solved the problem that he wanted to share the accomplishment with others. Interpreted in this way, Ronnie's intent was not to violate the classroom norms that obliged children to figure out problems for themselves; consequently, a negative sanction, such as expression of anger, against Ronnie was unwarranted.

Thus far, we have focused on reinterpretations initiated by the teacher. The children also responded to the emotional acts of their partners by suggesting alternative construals and, in doing so, sustained the classroom social norms. The following episode is taken from a class period in which the children solved story problems in small groups. After 20 minutes of intense effort, Andy and Rodney were still working on their first problem.

| Rodney: | I've been sitting here all day not figuring out this problem. Look at John and them. They're on their third problem. |
|---|---|
| Andy: | They didn't get this one. |

Andy's remark expressed the belief that persistence and figuring out personally challenging problems were expected in this class, and that fulfilling this obligation was more important than completing as many problems as possible. His statement implicitly indicated to Rodney that he should not interpret the situation as warranting negative emotions such as frustration or anxiety.

Another example from small group work early in the year shows how a task-motivated child, Kara, responded to an ego-motivated child, Lois, by suggesting an alternative interpretation of a situation. Kara and Lois had just completed the following problem.

4    6                                                              ↓
·    ·    ·    ·    ·    ·    ·    ·    ·    —

The problem is to determine the number indicated by the arrow and to indicate at the right a relationship between successive numbers (e.g., plus 2, add 2, or skip 1 number). The children's paper looked like this:

4    6    8    10    12    14    16    18
                                     ↓
·    ·    ·    ·    ·    ·    ·    ·    ·    —

Lois:    Erase those [numbers above the intermediate dots] because
         then they can't tell that you did that.
Kara:    We should leave them in because they help you.

Lois' comment indicates the public self-awareness that is typical of ego-involvement (Nicholls, 1983) because she would be embarrassed if others ("they") saw that she and Kara had used what she considered to be a relatively immature method. Such embarrassment would be appropriate in a situation in which children are publicly compared with others on the basis of their solution methods. Kara's comment implies that these concerns were misplaced in this classroom and that whatever method they used to solve a problem was acceptable. With regard to Kara's interpretation, embarrassment was not warranted in the event that others saw their paper.

## Beliefs About the Teacher's Role

Unlike traditional mathematics classes in which children typically experience frustration when they encounter situations in which they do know what to do (McLeod, this volume, Chapter 2), children in the project classroom quickly learned that not knowing what to do was routine. Also, the process of genuine problem solving became an overriding feature of mathematical activity. The children's understanding of the teacher's role as one of framing problematic situations and facilitating solution processes developed simultaneously with their beliefs about the nature of mathematical activity. The teacher frequently responded to students' questions about what they were "supposed to do" with "I don't know" or "You are going to have to figure that out for yourself." Students accepted the teacher's response from the outset but interpreted it differently as the year progressed. To illustrate the student's evolving beliefs about the teacher's role, we first consider an episode that occurred early in the year. One pair of children was working on an arithmetical task in which they had to decide what number to put in an empty box to equilibrate a pan balance.

> Ann:    I'm going to ask Mrs. M if we're supposed to add or subtract. (Ann goes off to talk to the teacher. She returns and reports to her partner.)
> Ann:    She doesn't know either.

Later in the year, Ann came to understand that the teacher's failure to tell her what to do was not an indication of ignorance, but instead implied that she expected the children to figure things out for themselves. This was indicated by Ann's recommendation to another child; "Don't ask her [Mrs. M.]. She won't tell you." Once the children had reconceptualized their understanding of the teacher's role, they had no reason to become angry or frustrated because she would not tell them what they were to do. At the same time, the children realized that they were not expected to use any particular method that the teacher had in mind.

## Beliefs About Doing Mathematics

At the beginning of the school year, the teacher guided the mutual construction of social norms that made it possible for the children to freely express their mathematical ideas for the remainder of the year. The following episode occurred during the third day of the school year. The discussion centered on the word problem "How many runners altogether? There are six runners on each team. There are two teams in the race."

> Teacher:    Jack. What answer-solution did you come up with?
> Jack:    14.
> Teacher:    14. How did you get that answer?
> Jack:    Because 6 plus 6 is 12. 2 runners on 2 teams . . . (Jack stops talking, puts his hands to the side of his face and looks down at the floor. Then he looks at the teacher and then at his partner, Ann. He turns and faces the front of the room with his back to the teacher, and mumbles incoherently.)
> Teacher:    Would you say that again. I didn't quite get the whole thing. You had . . . Say it again, please.
> Jack:    (Softly, still facing the front of the room.) It's 6 runners on each team.
> Teacher:    Right.
> Jack:    (Turns to look at the teacher.) I made a mistake. It's wrong. It should be 12. (He turns around and faces the front of the room.)

The teacher realized that Jack had interpreted the situation as warranting acute embarrassment. His concern with social comparison confounded the teacher's desire that the children should feel free to publicly express their thinking. For her purposes, it was vital that children feel no shame or embarrassment when they present solutions in front of others. Consequently, she immediately responded:

| | |
|---|---|
| Teacher: | (Softly.) Oh, okay. Is it okay to make a mistake? |
| Andrew: | Yes. |
| Teacher: | Is it okay to make a mistake, Jack? |
| Jack: | (Still facing the front of the class.) Yes. |
| Teacher: | You bet it is. As long as you're in my class it is okay to make a mistake. Because I make them all the time, and we learn from our mistakes, a lot. Jack already figured out, "Oops. I didn't have the right answer the first time" (Jack turns and looks at the teacher and smiles), but he kept working at it and he got it. |

Later in the year, as the following episode illustrates, admissions of mistakes were no longer construed as warranting embarrassment or shame but were instead viewed simply as events that occur in the normal course of classroom life.

| | |
|---|---|
| Charles: | 67. |
| Teacher: | 67. (She starts to write the answer.) |
| Joel: | Disagree. |
| Teacher: | All right, Joel. What do you think? |
| Joel: | 72. |
| Teacher: | You think it's 72 (several students disagree). |
| Joel: | Well . . . (he stands up and walks to the front of the class). |
| Teacher: | Let's listen to Joel's explanation. |
| Joel: | (Stands looking at the board.) I used 25 and 10 makes 35. And another 10 makes 45 and another 10 makes 55. (He stops and looks at his paper in his hand.) Another makes 65 (pause) [and 2 more make] 67. (He turns and looks at the teacher.) I disagree with my answer. (He smiles.) |
| Teacher: | (Laughing) I like that. I disagree with my answer. That's great. Raise your hand if you ever disagree with your own answer. It happens to all of us. |

Because children came to believe that doing mathematics is essentially a problem-solving activity, typical negative emotions such as anxiety, embarrassment, and shame, which accompany the obligation of producing publicly evaluated solutions to a large number of tasks in a quick, error-free manner by using prescribed methods, did not occur in this classroom. As in any classroom, however, occasions when a child transgressed the social norms were construed by others as situations warranting negative emotional acts. For example, because the children felt obliged to figure things out and to be able to justify their answers, they did exhibit frustration, disappointment, and sometimes anger with their peers when they were denied the opportunity to do just that. In these cases, the negative emotional acts were directed at other children and not at mathematics or the teacher. The following episode illustrates this point. Connie and Rodney worked on the problem of determining which number goes in place of the arrow in the following number-dot sequence:

$$3 \qquad 8 \qquad 13 \qquad\qquad\qquad \downarrow$$

.     .     .     .     .     .     .     .     —

Rodney had figured out that 18 should go above the first dot after 13, but Connie did not understand how he arrived at this result. Ann, a member of a neighboring pair, leaned over to tell them the answer that she and her partner had came up with.

| | |
|---|---|
| Ann: | The answer is 21. |
| Connie: | (To Rodney) The answer is 21. |
| Rodney: | (Angrily) No it isn't. This is—look! I'm going to figure it my *own* way. |
| Connie: | I already told you. |
| Rodney: | I don't want to copy. |
| | (At this point, the teacher comes by to observe the group working.) |
| Rodney: | (To the teacher) I don't want them to tell me the answer! |
| Teacher: | They want to—I know you're trying to work— |
| Rodney: | (Interrupting her) I want to try to work it out myself, but they're over here telling me the answer and everything. I think it should be 18 because . . . (Rodney explains his thinking to the teacher.) |

The anger Rodney displayed in this episode was appropriate according to the norms established in this classroom. Children were expected to construct justifiable solutions and not simply fill in answers. Rodney's display of anger sustained those norms and served to remind both Ann and Connie that their actions had deprived him of the opportunity to fulfill his obligations. The teacher's affirmation of the rationale for his anger gave his interpretation of the situation further credence.

Later in the year, the following incident occurred during a whole-class discussion. The episode began as the teacher called on Dan and his partner Brenda for a solution.

| | |
|---|---|
| Teacher: | Dan. |
| Dan: | They [another pair] were bothering us. |
| Brenda: | They were telling us the answers. |
| Teacher: | Oh. . . . You know when people give you the answers, boys and girls, does that really help you understand what you're doing? You don't know how you got it, you might as well just not waste your pencil. |

The teacher recognized the appropriateness of Dan and Brenda's complaint and immediately initiated a conversation in which she reminded the children that understanding what they were doing mathematically was paramount.

Teacher:     If you don't know what you are doing, it isn't going to help you get the answer. It's like saying, "Yup the answer is 7. Yes, I got it right. How did I get 7? I *don't* know." That doesn't help you one single bit. I know you are all friends. I know you want to help each other, but you help each other more by helping each other figure out an answer, rather than saying "7. Just write down 7. That's the answer. Trust me." You have to try and understand it.

By explicitly telling the children that it was wrong to give others the answer, the teacher used the incident to indicate to the rest of the class that Dan's and Brenda's indignation was warranted. She further sustained the social norm by posing a problematic situation for Dan and Brenda to solve.

Teacher:     How did you handle the situation?
Brenda:      We just said we didn't want the answer.... We were on the same question, and they were telling us the answer. We didn't pay any attention, because we wanted to figure it out for ourselves.
Teacher:     Good! Good for you. I'm proud of you. It's easy to take someone else's answer, isn't it, than to think about it yourself?
Students:    Yeah.
Teacher:     Sure it is. I could just fill these all in (points to the problems) and say, "These are the answers, kids." But would you learn anything?
Brenda:      (Interrupts) We have to think for ourselves. We can't have other people think for us....They might be wrong.
Teacher:     That's right.

## Positive Emotional Acts

The project classroom was characterized by both a general absence of negative emotional acts and the frequent occurrence of positive emotional acts when solving mathematical (as opposed to social) problems. Visitors to the project classroom invariably remarked about the excitement for mathematics displayed by the children as they solved the activities. Children frequently jumped up and down, hugged each other, and rushed off to tell the teacher when they solved a particularly challenging problem. Significantly, the positive emotional acts occurred when the children completed personally challenging tasks or constructed mathematical relationships. Because doing mathematics is thought by many, including many mathematics educators, to be associated with negative emotion (McLeod, 1985), it is especially important to clarify that, in the project classroom, positive emotional acts were not reactions to extraneous factors, such as receiving extrinsic rewards or ego satisfaction, but stemmed directly from mathematical activity. In this sense, the emotional acts of the children parallel those of the mathematician when solving a problem or developing an elegant proof (Silver & Metzger, this volume, Chapter 5). To illustrate, we present examples from both whole-class discussions and small-group work.

The whole-class episode began with a problem from an activity called *number-line.*

$$3 \qquad 8 \qquad 13 \qquad\qquad \downarrow$$

.    .    .    .    .    .    .    .    —

The children were to figure out what number goes above the dot indicated by the arrow and to construct a relationship between successive numbers.

| | |
|---|---|
| Teacher: | We have a 3, 8, 13, and then nothing. Ann, how did you and Alex do this? |
| Ann: | We got 25. |
| Teacher: | Okay. This is 25. (On an overhead transparency, she writes 25 over the dot where the arrow is pointing.) |
| Alex: | The pattern was add 5. |
| Teacher: | Plus 5. Good. I'm going to keep going very quickly. If you disagree, shoot up your hand and say, "I disagree." (Several students raise their hands and say disagree.) Oh, my gosh! Good thing I stopped when I did. Okay, Jeff what did you want to say? |
| Jeff: | It should be 22. |
| Teacher: | 22. (She writes 22 under the 25. Dan, Brenda, Lisa, and Johanna wave their hands frantically in the air saying "No, Uh! Uh!" John turns and looks at Jeff and shakes his head no.) |
| Teacher: | Gee. Here we go (laughs). Dan, What do you say? |
| Dan: | 18. |
| Teacher: | You say 18. Okay. (She writes 18 under 22 and pauses. Hands wave in the air.) Kirsten. |
| John: | (Speaks out.) He wants to change his answer. |
| Peter: | (Shouts out.) I disagree! |
| Teacher: | I know you do. I hear you loud and clear. |
| Lisa: | (Standing up waving her hand frantically.) It's 23! It's 23! |

The children continued giving their answers. Finally, the teacher asked:

| | |
|---|---|
| Teacher: | How are we going to figure this out very quickly? Lisa? |
| Lisa: | Count on our hands, 13, 14, . . . 23. |

| Teacher: | What's another way of checking your work? |
| --- | --- |
| Andrew: | Well, the pattern is plus 5. (He stands up and rushes excitedly to the front of the class and gestures at the screen.) This dot is 13. (He counts the next two dots.) 5 plus 5 is 10. Just add 10 to 13. |
| Teacher: | 13 plus 10 is . . . |
| Students: | (In unison) 23. |
| Teacher: | 23. You bet. |

As the relative merits of solutions were never discussed in the whole-class setting, it is unlikely that Andrew was excited because his solution was the most sophisticated. Rather, he had reconceptualized his understanding of the task and construed his insight as warranting excitement. Furthermore, with respect to the classroom social norms, he gave his explanation to share this insight, not to show how clever he was. The other children seemed genuinely pleased by his breakthrough.

The following dialogue once again illustrates the excitement that the children typically experienced when they constructed mathematical relationships. The dialogue is between two children as they solved a sequence of multiplication tasks corresponding to the sentences $10 \times 4 = $ ____, $9 \times 4 = $ ____, $8 \times 4 = $ ____, and $8 \times 5 = $ ____. (The children's use of the term *sets* in talking about multiplication derives from the teacher's use of the term when she first introduced "×" as the mathematical symbol for multiplication.) They are working on $10 \times 4 = $ ____ after having found $5 \times 4 = 20$.

| John: | It's five more sets [of 4]. Look. Five more sets than 20. |
| --- | --- |
| Andy: | Oh! 20 plus 20 is 40. So its gotta be 40. No. |
| John: | Yeah! |
| Andy: | No. 4, 8, 12, 16, 20, 24, 28, . . . (keeping track on his fingers). |
| John: | 40. |
| Andy: | 40. |
| John: | Yeah, I know . . . 'cause ten 4s make 40. |
| Andy: | Like five 4s make 20. |
| John: | Four sets of 10 makes 40. Just turn it around. |
| Andy: | Five sets of 4s make 20 and so five more than that. |
| John: | Yeah, just turn it around. Just turn it around. |
| Andy: | 5 times 4 is 20, so 20 more than that makes 40. |
| John: | Just switch them around. |

John's visible excitement did not stem from the fact that he had solved the problem because he had previously arrived at 40 as an answer by relating $10 \times 4 = $ ____ to $5 \times 4 = 20$. Instead, it derived from his construction of the principle of commutativity of multiplication, which allowed him to develop a second, more satisfying solution to the problem. John's reaction was analogous to that of the mathematician who succeeds in proving a theorem by a particularly elegant method. At this point in the episode, Andy did not display any emotion. John's repeated comment, "Just turn it around," is an indication that he is trying to convey both his excitement and his insight to Andy, but Andy is oblivious to John's intent. As the episode continued, John again exhibited excitement when he constructed a relationship between successive tasks.

$9 \times 4 =$ ____

> John:    Just take away 4 from that [$10 \times 40$].
> Andy:    36.
> John:    (pause) Yeah!

$8 \times 4 =$ ____

> John:    Look! Look! Just take away 4 from that [$9 \times 4$] to get that [$8 \times 4$]. See! Just take away 4 from there [$9 \times 4$]. (The dialogue continues as Andy solves the problem using a different method.)

$8 \times 5 =$ ____

> Andy:    Five more than that [$8 \times 4$] is 37.
> John:    Eight sets of 4. Eight sets of 5.
> Andy:    No. 9, 39, I think.
>          (Both children pause to reflect for a few moments.)
> John:    (Very excitedly) It's 40.
> Andy:    It is?
> John:    Yeah, it's 40! Yeah, look!

The episode concluded as John demonstrated his method and Andy verified it. Throughout, John repeatedly displayed excitement at having constructed mathematical relationships. It was his construction of mathematical knowledge, in and of itself, that gave rise to this excitement.

## Reflections

In presenting the sample episodes from the classroom we have attempted to illustrate that children's beliefs, their emotional acts, and the network of obligations and expectations that constitute the social context within which they do mathematics are all intimately related. Consider, for example, beliefs and emotional acts. The process of attributing a particular emotional quality to someone's actions necessarily involves inferences about his or her construal of the situation. This construal in turn reflects underlying beliefs. Consequently, emotional acts can be viewed as expressions of beliefs and, from the observer's perspective, are valuable sources of insight into the possible nature of those beliefs (Cobb, 1986c). In other words, if someone acts in an emotional way, we know that they really mean it, and a "willingness to act and ... the assumption of some risk and responsibility for action in relation to a belief represent essential indices of actual believing" (Smith, 1978, p. 24). In short, emotional acts depend on beliefs.

The converse, that beliefs depend on emotional acts, is at least partially true. We have seen how the teacher frequently capitalized on the children's emotional acts to renegotiate the classroom social norms. In doing so, she explicitly discussed obligations that she expected the children to fulfill. To the extent that the children accepted these obligations—and there is every indication that they did, for the most part—they reorganized their beliefs about their own role, the teacher's role, and the activity of doing mathematics. Thus, the teacher intuitively agreed with Smith's dictum and spontaneously focused on the children's emotional acts as prime indicators of their beliefs. Within the unfolding stream of classroom life, the teacher acted on the basis of her interpretations of the children's emotional acts and gave the children opportunities to reorganize their beliefs. We use the phrase "gave them opportunities" for the simple reason that beliefs can no more be transmitted from one person to another than can conceptual knowledge of mathematics. The teacher helped the children construct their own beliefs by, in effect, setting puzzles for them to solve. For example, she told the children in no uncertain terms that the phrase "that's easy" was inappropriate. In the last analysis, however, the children had to figure out for themselves why it was inappropriate. They had to understand that this could make other people feel bad, and that making people feel bad is morally wrong. Their ability to reorganize their beliefs in a way that was compatible with the teacher's expectations depended on their current conceptions of morality—that is, on how they interpreted the teacher's statements. We note in passing that the puzzles set by the teacher also gave opportunities for moral growth and, thus, the development of moral autonomy. This was one of the initial noncognitive goals of the project.

It has been argued that beliefs span the cognitive and affective domains (Schoenfeld, 1985). Our analysis expresses the alternative view that beliefs are basically cognitive but that they function in the construals that generate emotional acts. We suggest that beliefs span the individual and social domains. We argued previously, for example, that one has to consider unfolding classroom social life in order to appreciate the partial dependence of beliefs on emotional acts. A child's beliefs about his or her own role and the teacher's role are, in fact, the child's understanding of the classroom social norms. It should be clear that here we include both the implicitly and the explicitly held beliefs that give rise to obligations and expectations in the course of social interactions in the classroom. In fact, to infer that a child holds particular beliefs is a way of summarizing the child's inferred obligations and expectations in a variety of concrete situations.

The claim that beliefs span the individual and the social domain is an instance of the more general contention that neither the individual nor the social domain is primary (Bauersfeld, 1988; Cobb, 1986c; Cobb, Wood, & Yackel, in press b; Voigt, 1985). One cannot adequately analyze one without considering the other, because the activities of individuals (including their emotional acts) serve to construct the social norms that constrain those very same activities. Conversely, the norms constrain the activities that construct the norms. Thus, to acknowledge that social context is an integral part of an individual's cognition and affect does not imply that social norms are taken as solid, independently existing bedrock upon which to anchor analyses of learning and teaching.

In the first section of this chapter, we proposed that children's beliefs about their own role, the teacher's role, and the nature of mathematical activity comprise a vital core of children's standards of normative expectancies. Emotional acts occur when experience is incompatible with these expectancies. At several points in the analysis of sample episodes, we hinted that the three types of beliefs develop together. In fact, it was for this reason that we avoided language that might suggest that they are three independent components. Instead, we view them as mutually dependent aspects of a self-organizing system. It is readily apparent that a child's beliefs about his or her own role and the teacher's role are intimately connected. In our project classroom, for example, the children's beliefs about the teacher's role changed with the realization that they were obliged to resolve their problems for themselves and that they were not obliged to use any particular solution method. Their beliefs about the nature of mathematical activity also changed once they accepted and attempted to fulfill this obligation. Similarly, the belief that mathematical solutions should be justifiable evolved together with beliefs about their own role and the teacher's role during both small-group work and whole-class discussions. Consideration of these relationships suggests a way to get at the system in belief systems. As we have seen, emotional acts played a crucial role in the children's construction of these systems.

Finally, the reader may recall that nurturing the development of the children's intellectual autonomy was a major noncognitive goal of the project. The astute reader may have noted that many of the sample episodes presented were strong indicators of intellectual autonomy. In the course of the teaching experiment, we came to realize that the manner in which the children reorganized their beliefs was synonymous with the growth of both intellectual and moral autonomy, developments that involved emotional acts. We now prefer to talk of beliefs rather than autonomy because the latter is a more global, less easily differentiable construct.

## Implications

The sample episodes serve as prototypical cases of the ways in which the teacher capitalized on the children's emotional acts to initiate, guide, and sustain the mutual construction of what Silver (1985) called a *problem-solving atmosphere*. The episodes exemplify our view of what constitutes effective mathematics teaching. More far-reaching implications become apparent when we relate our analysis of the project classroom to current theories of achievement motivation, particularly that of Nicholls (1983, 1984, 1987).

Nicholls distinguishes between two conceptions of ability. The more differentiated conception is that embodied in standard ability-testing procedures, in which ability is defined with reference to the performance of others. When this conception of ability is operative, effort is considered to improve mathematical learning and problem solving only up to the limit of one's present capacity; "that is, ability is conceived as capacity—an underlying trait that is not observed directly but is inferred from both effort and performance, in a context of social comparison" (Nicholls, 1984, p. 41). When two students are equally successful in solving mathematical problems, for example, but one had to expend much greater effort than the other, we would conclude with respect to the differentiated conception that the more conscientious student was less able. Students who assess their performance in terms of this conception and believe that their ability is low come to believe that they lack capacity. They believe that persisting and doing their best will frequently not be good enough because of inadequacies that are beyond their control.

When the undifferentiated conception of ability is operative, "high ability is implied by learning or by success at tasks they [students] are uncertain of being able to complete. They do not judge ability with reference to performance norms or social comparison" (Nicholls, 1984, p. 41). A sense of being able comes from persisting in solving a personally challenging problem. The fact that others might have solved the problem with less effort is irrelevant. As Nicholls (1984, p. 42) stated: "Ability does not, in this case, imply an inferred trait . . . When more effort is needed for success, this implies more learning, which is more ability . . . The subjective experience of gaining insight or mastery through effort is the experience of competence of ability." Individuals do not typically assess their performance uniformly in terms of one or the other conception of ability; for example, a student might typically employ the differentiated conception during mathematics lessons and the undifferentiated conception in art lessons. In other words, use of the conceptions is contextual.

The promotion of task-involvement as a form of motivation (and thus employment of the less differentiated conception of ability) was the third major noncognitive goal of the project. We have provided initial documentation elsewhere (Cobb, Wood, & Yackel, in press a) that the project was extremely successful in this regard. Furthermore, the emotional acts presented in the sample episodes are consistent with the inference that the children became increasingly involved in their mathematical problems; for example, they became excited and reported that they felt good when they solved personally challenging problems, irrespective of whether other groups had already completed the same instructional activities. Negative emotional acts occurred when they were deprived of the opportunity to think things through for themselves, as when another child told them the answer, but not when they struggled to solve a problem. This success is directly attributable to the manner in which the teacher initiated and guided the mutual construction of the classroom social norms. Thus, the children's reorganization of their beliefs occurred concomitantly with the development of task-involvement as a form of motivation. (See Silver, 1982, and Cobb, 1985, for more detailed discussions of the relationship between beliefs and motivations.)

The most general implication of our work is that the teacher should renegotiate the social context within which children attempt to solve mathematical problems and thus influence their beliefs about their own and the teacher's roles and the nature of mathematical activity. The objective is for both the teacher and the students to create a social context in which construals that warrant detrimental negative emotions such as frustration are simply not made while solving mathematical problems. This recommendation can be contrasted with those that attempt to help students cope with emotions that are warranted in ego-involving situations.

Consider, for example, recommendations derived from Weiner's (1979) casual attribution theory. Weiner conducted an analysis of the possible causes of (what he takes for granted is) success and failure and classified them according to whether the causes are internal or external to the individual, stable or unstable, and controllable or uncontrollable. Within this scheme, ability is considered to be an uncontrollable, stable, internal cause. This is the differentiated conception of ability—ability as capacity. With regard to this conception, "if students attribute their failures in problem solving to their lack of ability, they are likely to be unwilling to persist in problem-solving tasks very long" (McLeod, 1985, p. 275). As long as the differentiated conception of ability is accepted unquestioningly as the way things are and must be, interventions are limited to persuading students to make alternative attributions. We, in contrast, suggest that the problem of deleterious attributions disappears if social norms are renegotiated to encourage task-involvement and the undifferentiated conception of ability. Students would then "have no reason to consider the role of factors such as ability, difficulty, or luck because effort can be perceived directly. . . . Thus, if effort attributions were the prime mediators of achievement affect, causal attribution theory would be irrelevant to achievement motivation" (Nicholls, 1984, p. 62).

To clarify this point, consider frustration as an emotional act. Frustration is generally warranted in situations in which one is unable to achieve one's purposes. In terms of Weiner's scheme, this emotion is appropriate when students attribute failure to lack of ability, in the differentiated sense, because it is beyond their control. In line with the claim that the children in the project classroom became increasingly task-involved during mathematics instruction, we were unable to identify a single instance during the second semester in which a child became frustrated and gave up because he or she could not complete an instructional activity. We observed children who, when compared with their peers, failed repeatedly day after day, yet these children continued to persist during small-group work, contributed to whole-class discussions, and achieved personal satisfaction by doing so. In terms of the undifferentiated conception of ability, these children were not failing. Their purpose was not to demonstrate superior capacity, avoid looking stupid, or put one over on the teacher. Engaging in mathematical activity was an end in itself, and as long as they did their best they considered themselves to be succeeding. Indeed, they were succeeding in a classroom in which the social norms obliged them to think their problems through for themselves and take responsibility for their own learning. With respect to these norms, then, frustration was not warranted.

In conclusion, we contend that well-meaning attempts to either persuade students to make alternative attributions of the cause of failure or to teach students about detrimental affective variables that might influence them miss an essential point. The students' undesirable attributions and affects are appropriate only with respect to the social norms established in settings that are characterized by competition, social comparison, and public self-awareness and, thus, induce ego-involvement (Nicholls, 1983). The same can be said of recommendations to develop blockage-free instruction or to design computer-based learning environments that alleviate blockages and thus reduce frustration. (We believe that such environments can have educational value. We are merely questioning one proposed rationale for their use.) Here again, it is assumed that blockages are inevitably construed as warranting negative emotions that interfere with learning. If one believes, as we do, that substantive mathematical learning is a problem-solving process, them attempts to exorcise or ameliorate blockages or problematic situations reduce students' opportunities to learn. Surely, the solution to the problem of students' negative emotions during mathematical problem solving is not to quash mathematical problem solving. We suggest instead that it is more productive to initiate and guide the construction of alternative social norms with respect to which deleterious emotions and attributes are not warranted. The sample episodes illustrate how one teacher capitalized on children's emotional acts to do just this and, in the process, achieved greater psychic rewards (Lortie, 1975) as she observed the children learn in her classroom.

*Acknowledgments.* The research reported in this paper was supported by the National Science Foundation under Grant No. MDR-8740400. All opinions and recommendations expressed are, of course, solely those of the authors and do not necessarily reflect the position of the Foundation.

## References

Armon-Jones, C. (1986a). The thesis of constructionism. In R. Harré (Ed.), *The social construction of emotions* (pp. 33–56). Oxford: Blackwell.

Armon-Jones, C. (1986b). The social functions of emotions. In R. Harré (Ed.), *The social construction of emotions* (pp. 57–82). Oxford: Blackwell.

Averill, J. (1986). The acquisition of emotions during adulthood. In R. Harré (Ed.), *The social construction of emotions* (pp. 98–118). Oxford: Blackwell.

Balacheff, N. (1986). Cognitive versus situational analysis of problem-solving behavior. *For the Learning of Mathematics*, 6(3), 10–12.

Bateson, G. (1973). *Steps to an ecology of mind.* London: Paladin.

Bauersfeld, H. (1988). Interaction, construction, and knowledge: Alternative perspectives for mathematics education. In T. Cooney & D. Grouws (Eds.), *Effective mathematics teaching* (pp. 27–46). Reston, VA: National Council of Teachers of Mathematics and Hillsdale, NJ: Lawrence Erlbaum Associates.

Beford, E. (1986). Emotions and statements about them. In R. Harré (Ed.), *The social construction of emotions* (pp. 15–31). Oxford: Blackwell.

Bishop, A. (1985). The social construction of meaning: A significant development for mathematics education? *For the Learning of Mathematics*, 5(1), 24–28.

Blumer, H. (1969). *Symbolic interactionism*. Englewood Cliffs, NJ: Prentice-Hall.

Brousseau, G. (1984). The crucial role of the didactical contract in the analysis and construction of situations in teaching and learning mathematics (Occasional paper 54). In H.G. Steiner (Ed.), *Theory of mathematics education* (pp. 110–119). Bielefeld, Germany: IDM.

Bruner, J. (1986). *Actual minds, possible worlds*. Cambridge, MA: Harvard University Press.

Cobb, P. (1985). Two children's anticipations, beliefs, and motivations. *Educational Studies in Mathematics, 16*, 111–126.

Cobb, P. (1986a). Clinical interviewing in the context of research programs. In G. Lappan & R. Even (Eds.), *Proceedings of the Eighth Annual Meeting of the North American Chapter of the International Group for the Psychology of Mathematics Education: Plenary speeches and symposium* (pp. 90–110). East Lansing, MI: Michigan State University.

Cobb, P. (1986b). Concrete can be abstract: A case study. *Educational Studies in Mathematics, 17*, 37–48.

Cobb, P. (1986c). Contexts, goals, beliefs, and learning mathematics. *For the Learning of Mathematics, 6*(2), 2–9.

Cobb, P., & Merkel, G. (in press). Thinking strategies as an example of teaching arithmetic through problems solving. In P. Trafton (Ed.), *1989 yearbook of the National Council of Teachers of Mathematics*. Reston, VA: National Council of Teachers of Mathematics.

Cobb, P., & Steffe, L.P. (1983). The constructivist researcher as teacher and model builder. *Journal for Research in Mathematics Education, 14*, 83–94.

Cobb, P., & von Glasersfeld, E. (1984). Piaget's scheme and constructivism. *Genetic Epistemology, 13*(2), 9–15.

Cobb, P., Wood, T., & Yackel, E. (in press a). Implications of the new philosophy of science for mathematics education. In J. Novak (Ed.), *The importance of the new philosophy of science for school instruction*. New York: Westside Press.

Cobb, P., Wood, T., & Yackel, E. (in press b). Learning through problem solving: A constructivist approach to second grade mathematics. In E. von Glasersfeld (Ed.), *Constructivism in mathematics education*. Holland: Reidel.

Confrey, J. (1984, April). *An examination of the conceptions of mathematics of young women in high school*. Paper presented at the annual meeting of the American Educational Research Association, New Orleans.

Confrey, J. (1987, July). *The current state of constructivist thought in mathematics education*. Paper presented at the annual meeting of the International Group for the Psychology of Mathematics Education, Montreal, Canada.

Coulter, J. (1986). Affect and social context: Emotion definition as a social task. In R. Harré (Ed.), *The social construction of emotion* (pp. 120–134). Oxford: Blackwell.

Davydov, V.V. (1975). The psychological characteristics of the "prenumerical" period of mathematics instruction. In L.P. Steffe (Ed.), *Soviet studies in the psychology of learning and teaching mathematics* (Vol. 7, pp. 109–205). Stanford, CA: School Mathematics Study Group.

Erickson, F. (1985). Qualitative methods in research on teaching. In M.C. Wittrock, (Ed.), *Handbook of research on teaching* (3rd ed., pp. 119–161). New York: Macmillan.

Goodlad, J.I. (1983). *A place called school: Prospects for the future*. New York: McGraw-Hill.

Hargreaves, D.H. (1975). *Interpersonal relations and education.* London: Routledge and Kegan Paul.

Harré, R. (1986). An outline of the social constructionist viewpoint. In R. Harré (Ed.), *The social construction of emotions* (pp. 2–14). Oxford: Blackwell.

Hiebert, J., & Lefevre, P. (1986). Conceptual and procedural knowledge in mathematics: An introductory analysis. In J. Hiebert (Ed.), *Procedural and conceptual knowledge: The case of mathematics* (pp. 1–27). Hillsdale, NJ: Lawrence Erlbaum Associates.

Hundeide, K. (1985). The tacit background of children's judgments. In J.V. Wertsch (Ed.), *Culture, communication, and cognition* (pp. 306–322). Cambridge, England: Cambridge University Press.

Kamii, C. (1985). *Young children reinvent arithmetic: Implications of Piaget's theory.* New York: Teachers College Press.

Labinowicz, E. (1985). *Learning from children.* Menlo Park, CA: Addison-Wesley.

Levina, R.E. (1981). L.S. Vygotsky's ideas about the planning function of speech in children. In J.V. Wertsch (Ed.), *The concept of activity in Soviet psychology* (pp. 279–299). Armonk, NY: Sharpe.

Lortie, D.C. (1975). *Schoolteacher.* Chicago: University of Chicago Press.

McLeod, D.B. (1985). Affective issues in research on teaching mathematical problem solving. In E.A. Silver (Ed.), *Teaching and learning mathematical problem solving: Multiple research perspectives* (pp. 267–280). Hillsdale, NJ: Lawrence Erlbaum Associates.

Mehan, H. (1979). *Learning lessons: Social organization in the classroom.* Cambridge, MA: Harvard University Press.

Menchinskaya, N.A. (1969). Fifty years of Soviet instructional psychology. In J. Kilpatrick & I. Wirszup (Eds.), *Soviet studies in the psychology of learning and teaching mathematics* (Vol. 1, pp. 5–27). Stanford, CA: School Mathematics Study Group.

Nicholls, J.G. (1983). Conceptions of ability and achievement motivation: A theory and its implications for education. In S.G. Paris, G.M. Olson, & W.H. Stevenson (Eds.), *Learning and motivation in the classroom* (pp. 211–237). Hillsdale, NJ: Lawrence Erlbaum Associates.

Nicholls, J.G. (1984). Conceptions of ability and achievement motivation. In R.E. Ames & C. Ames (Eds.), *Research on motivation in education: Vol.1, Student motivation* (pp. 39–73). New York: Academic Press.

Nicholls, J.G. (1987). *Motivation, values, and education.* Paper presented at the annual meeting of the American Educational Research Association, Washington, DC.

Perret-Clermont, A.N. (1980). *Social interaction and cognitive development in children.* New York: Academic Press.

Piaget, J. (1970). *Genetic epistemology.* New York: Columbia University Press.

Piaget, J. (1980). *Adaptation and intelligence: Organic selection and phenocopy.* Chicago: University of Chicago Press.

Pritchard, M. (1976). On taking emotions seriously. *Journal for the Theory of Social Behavior, 6*(2), 1–27.

Rumelhart, D.E., & Norman, D.A. (1981). Analogical processes in learning. In J.R. Anderson (Ed.), *Cognitive skills and their acquisition* (pp. 335–359). Hillsdale, NJ: Lawrence Erlbaum Associates.

Sarbin, T.R. (1986). Emotion and act: Roles and rhetoric. In R. Harré (Ed.), *The social construction of emotion* (pp. 83–97). Oxford: Blackwell.

Schoenfeld, A.H. (1985). *Mathematical problem solving.* New York: Academic Press.

Sigel, I.G. (1981). Social experience in the development of representational thought: Distancing theory. In I.E. Sigel, D.M. Brodzinsky, & R.M. Golinkoff (Eds.), *New directions in Piagetian theory and practice* (pp. 203–217). Hillsdale, NJ: Lawrence Erlbaum Associates.

Silver, E.A. (1982). Knowledge organization and mathematical problem solving. In F.K. Lester and J. Garofalo (Eds.), *Mathematical problem solving: Issues in research* (pp. 15–25). Philadelphia: Franklin Institute Press.

Silver, E.A. (1985). Research on teaching mathematical problem solving: Some underrepresented themes and needed directions. In E.A. Silver (Ed.), *Teaching and learning mathematical problem solving: Multiple research perspectives* (pp. 247–266). Hillsdale, NJ: Lawrence Erlbaum Associates.

Smith, J.E. (1978). *Purpose and thought: The meaning of pragmatism.* Chicago: University of Chicago Press.

Steffe, L.P. (1983). The teaching experiment methodology in a constructivist research program. In M. Zweng, T. Green, J. Kilpatrick, H. Pollack, & M. Suydam (Eds.), *Proceedings of the Fourth International Congress on Mathematical Education.* Boston: Birkhauser.

Steffe, L.P., Cobb, P., & von Glasersfeld, E. (1988). *Young children's construction of arithmetical meanings and strategies.* New York: Springer-Verlag.

Steffe, L.P., von Glasersfeld, E., Richards, J., & Cobb, P. (1983). *Children's counting types: Philosophy, theory, and application.* New York: Praeger Scientific.

Thompson, P. (1985). Experience, problem solving, and learning mathematics: Considerations in developing mathematics curricula. In E.A. Silver (Ed.), *Teaching and learning mathematical problem solving: Multiple research perspectives* (pp. 189–236). Hillsdale, NJ: Lawrence Erlbaum Associates.

Voigt, J. (1985). Patterns and routines in classroom interaction. *Recherches en Didactique des Mathématiques, 6,* 69–118.

von Glasersfeld, E. (1983). Learning as a constructive activity. In N. Herscovics & J.C. Bergeron (Eds.), *Proceedings of the Fifth Annual Meeting of the North American Chapter of the International Group for the Psychology of Mathematics Education* (Vol.1, pp. 41–69). Montreal: University of Montreal.

von Glasersfeld, E. (1984). An introduction to radical constructivism. In P. Watzlawick (Ed.), *The invented reality* (pp. 17–40). New York: Norton.

Weiner, B. (1979). A theory of motivation for some classroom experiences. *Journal of Educational Psychology, 71,* 3–25.

Wood, T., Cobb, P., & Yackel, E. (1988, April). *The influence of change in teacher's beliefs about mathematics instruction on reading instruction.* Paper presented at the annual meeting of the American Educational Research Association, New Orleans.

Wood, T., & Yackel, E. (1988, July). *Teacher's role in the development of collaborative dialogue within small group interactions.* Paper presented at the Sixth International Congress on Mathematical Education, Budapest, Hungary.

# 10
# Teaching Practices and Student Affect in Problem-Solving Lessons of Select Junior-High Mathematics Teachers

DOUGLAS A. GROUWS and KATHLEEN CRAMER

Affective variables are important in mathematical problem solving. For some people, they are important for their own sake, and for others they are important primarily because they facilitate productive problem-solving behavior. Regardless of one's perspective, however, it is the case that, as McLeod (1985) has observed, research in the mathematical problem-solving domain has concentrated on cognitive, rather than affective, issues. Another neglected theme in this research area is the role of the teacher in facilitating the development of problem-solving outcomes in classroom situations (Grouws, 1985). Good and Biddle (1988) have argued persuasively for including classroom observation in studies designed to increase our comprehension of mathematics teaching. The study reported here focuses on the teacher and affective issues, using classroom observation. In particular, the research explores the relationship between classroom structures and teaching behaviors and aspects of affect.

The general strategy of the study was to identify a small group of mathematics teachers who seemed to excel in the teaching of problem solving and then observe them as they taught problem-solving lessons. Particular attention was given during the observations to teaching behavior, classroom and lesson organization, and students' affective responses.

## Background Information

For the study, six teachers from a large urban school district were selected from a large pool of teachers who were identified by chairpersons of mathematics departments and the district mathematics coordinator as teachers who were doing excellent things with problem solving. They were successful in the sense that they met the following criteria: (a) they regularly taught problem solving and got good results; (b) they used innovative approaches to problem solving; and (c) they had some training in teaching problem solving. The teachers were from five different junior high schools. Four of these schools had a traditional configuration with students in two grade levels, seventh and eighth. These four schools also had similar socioeconomic make-ups. The fifth school, a K–8 alternative school, had an open program using multigrade classrooms and its students came from families with socioeconomic status higher than those of students in the other four schools.

The teachers differed in the amount of experience they had, the type of problems they used, and their instructional approaches to problem solving. The years of teaching experience ranged from 2 to 21 years. The types of problems varied from the familiar process-type problems found in commercial problem-solving supplementary materials to application problems. The process-type problems were solved using heuristics such as guess and check, look for a pattern, or solve a simpler problem. The applied problems were used to integrate problem solving with particular mathematics topics. The following problem, for example, was presented to a class after work on area and volume.

The MT Wastebasket Company wants to make a wastebasket rectangular in shape. The company wants the bottom to measure 8 inches by 10 inches. The total outside area is to be 656 square inches. What is the height?

Computers were used by three of the teachers as a means of providing problem-solving activities. The computer was utilized by students to either do programming using LOGO or to solve problems from a problem-solving software package.

In their instructional approaches, three teachers relied solely on group work with little direct instruction; others varied their approaches. For example, group work may have dominated the lesson one time, whereas direct instruction was used to teach a specific strategy at another time. One teacher consistently used a large-group direct-instruction model to motivate students and to help them understand the problem so that they would be successful when working in groups.

Despite the differences in experience, classroom organization, and types of problems, all six teachers conducted successful problem-solving lessons— successful in the sense that most students were engaging in higher order thinking and acquiring problem-solving experience. This chapter describes the positive affective characteristics of these teachers' classes and suggests hypotheses about how the instructional methods of teachers contributed to the positive student behaviors observed.

## Affective Characteristics of the Classes

With the exception of one teacher who was observed only three times because a student teacher joined her at mid-semester, each teacher was observed teaching five to seven problem-solving lessons over a 4-month period. Data collected on the affective characteristics of the classes represent only one facet of the information collected. The teacher was a primary focal point of the observation. The basic strategy for collecting the data was to record field notes on what the teacher did and said during the class, how the class was organized, and how students reacted. These notes were transcribed and summarized by completing an observation form for each lesson. An example of an observation form is presented in Fig. 10.1. The observational data were supplemented with an individual interview at the end of the school year.

The observations did not provide direct data on affective characteristics of individual students; hence, specific conclusions about the confidence and attitude or beliefs toward problem solving of individual students cannot be drawn. Comments can be made on class reactions to the problem-solving lessons that relate to three dimensions of the affective domain: willingness to work on problem-solving tasks, perseverance to complete tasks, and enjoyment in doing these tasks.

In all the classes, students showed a marked willingness to work on problems, they persisted to complete tasks correctly, and they seemed to enjoy the problem-solving lessons. In one class, for example, the teacher put commercially developed problem-solving cards on the chalk tray and had a representative from each team select a card. The students rushed to the board to get the more difficult cards, which were worth the most points. In this class, students were given feedback when they asked if their answers were correct. Consistently, students who were initially incorrect were observed trying to solve the problem a second time. In another class, students were observed to be so interested in the problem-solving lesson that they did not want to leave class at the bell: "Give us 5 more seconds. Oh! We're really close. We don't care if we are late for the next class."

One teacher used the force of his personality to motivate the students to solve the problems. This dynamic teacher used humor and a bit of theatrics to get students involved and interested and to increase their confidence so that they would be able to solve the problems.

Interviews with teachers provided additional information on these affective issues. When asked about how they perceived students' reactions to their problem-solving lessons, all of the teachers stated that the students looked forward to problem solving. One teacher commented that students did not ask for more worksheets on fractions but that they did ask for more problems. Other quotes from the interviews describe how teachers perceived their students' reactions to these classes: "More interested," "Energy level rises," "Pay more attention to what I say," "More energy and more fun," "With problem solving they don't complain that they have too much work," "More independent," "More excited and challenged," "Less discipline problems."

At this point, only hypotheses can be offered to explain why students in these classes exhibited these positive affective responses to problem-solving lessons. As a result of examining the dynamics of these classrooms, we believe that the following teaching practices may account for students' willingness to solve problems, their persistence in solving them, and their enjoyment of the lessons.

**Teacher Observation of Problem-Solving Lesson**

Name ___ Observation # ___ Date _____

School ___ Grade and level ___ # of students ___

Lesson Objective: _____

I. How many minutes are spent in these activities? (Also note by number the order in which they occur.)

____ Review

____ Homework

____ Mental computation

____ Quiz

____ Development of new lesson

____ Seatwork

____ Game

____ Directions

____ Non-math-related tasks

____ Other

II. Nature of development (note type and describe)

Lecture

Discovery

Question/answer

III. Teaching of problem solving

1. Organization of room
2. Grouping patterns (large, small, heterogeneous, cooperative, etc.)
3. Teacher's questions/comments to introduce problem/ activity (process vs. procedural)
4. Types of questions/comments to help students solve the problem (are specific strategies suggested?)
5. Ways the teacher models problem solving
6. Role of the answer in the lesson/activity
7. How teacher extends the problem/activity

FIGURE 10.1. Teacher observation form.

IV. Summary statements

    1. Amount of problem solving in class period

    2. Materials used (text, supplementary materials, manipulatives)

    3. Other indications of problem solving (POW, bulletin board)

    4. Types of problems used (one-step, two-step, process, real-life)

    5. Classroom climate (general atmosphere, student interest)

    6. Quality of lesson (5 4 3 2 1)

    7. Quotes about math

    8. Affective comments

High Inference Measures

Teacher ___ School ___ Observation # ___ Date _____

| Teacher's High Inference Measures | High | | Middle | | Low |
|---|---|---|---|---|---|
| 1. Degree to which presentations are meaningful | 5 | 4 | 3 | 2 | 1 |
| 2. Accomplishment | 5 | 4 | 3 | 2 | 1 |
| 3. Degree to which students are held responsible | 5 | 4 | 3 | 2 | 1 |
| 4. Managerial | 5 | 4 | 3 | 2 | 1 |
| 5. Teacher emphasis on higher order thinking | 5 | 4 | 3 | 2 | 1 |
| 6. Teacher use of manipulatives | 5 | 4 | 3 | 2 | 1 |

Group's High Inference Measures (Without Teacher)

| | High | | Middle | | Low |
|---|---|---|---|---|---|
| 1. Amount of time group is on task | 5 | 4 | 3 | 2 | 1 |
| 2. Amount of group interaction | 5 | 4 | 3 | 2 | 1 |
| 3. Higher cognitive level student behavior | 5 | 4 | 3 | 2 | 1 |
| 4. Amount of student cooperation | 5 | 4 | 3 | 2 | 1 |
| 5. Student use of manipulatives | 5 | 4 | 3 | 2 | 1 |
| 6. Group self-management | 5 | 4 | 3 | 2 | 1 |

FIGURE 10.1. (Continued).

## Dynamics of the Classrooms

All the classes were orderly and businesslike, with an obvious lack of discipline problems. Students were in class at the bell. Two teachers gave students a maintenance problem to complete during the first 5 minutes of class and had a process for quickly correcting homework and recording scores. In other classes, students were observed to wait quietly as the teacher took attendance. After roll was taken, they generally reviewed homework from the previous day or started immediately on the day's activity.

## *Teaching Rapport with Students*

Within this classroom order, there was a warm, friendly atmosphere. Students felt comfortable to joke a bit with the teachers. The teachers often shared their enthusiasm for the day's lesson. One teacher, for example, explained to the students that she had found one of the problems for that day on a poster that was hanging in a classroom she was in for a workshop. She said she became so involved in the problem that she did not listen to the speaker as she should have. Another teacher's humorous presentation of problems pleased his rather sophisticated audience; they would joke back to him about the "corniness" of his humor.

One teacher in the project felt comfortable enough with her class to confide that she was nervous about trying to do problem-solving lessons with her classes this year. She explained that last year's eighth-grade classes exhausted her and that she was skeptical about trying open-ended activities this year. She also complimented students on how well they worked during these lessons.

In general, there were clear indications of positive feelings between students and teachers. The teachers seemed happy to be teaching; the students seemed glad to be in these classes.

## *Accountability*

Accountability was another characteristic of each class. Although methods varied, students were held responsible for their work during problem-solving lessons. Their assignments were not extra credit; students received feedback on these assignments as they would on any other assignment. The following three strategies for evaluating students' problem-solving performance show the variability in how teachers held students accountable for their work. The factor that varied was the role of the answer.

### Strategy One

Figure 10.2 shows a record sheet used by students working on problem-solving task cards. The numbers down the left side are card numbers. Students worked in cooperative pairs on one card at a time, receiving feedback on the accuracy of their solution (and help, if incorrect) from the teacher before going on to the next

Names _____

_____

13) _____

14) _____          **Total Score**

15) _____     **blue**

51) _____     **× 2**

52) _____     ☐

53) _____


29) _____

30) _____

31) _____     **white**

32) _____     **× 3**

33) _____     ☐

34) _____


48) _____

49) _____

50) _____     **red**

86) _____     **× 4**

87) _____     ☐

88) _____

FIGURE 10.2. Record sheet for use with problem-solving cards.

card. Each card was assigned a point value depending on its difficulty. At the end of class, each pair of students tallied the points and handed in one record sheet. With this method, only the answer was evaluated.

STRATEGY TWO

In one class, students worked individually or in groups on a small set of problems. This teacher stressed the importance of being able to articulate clearly the solution and the strategy used to solve a problem. For example, the following problem was presented and ideas for solving it were discussed in a large group. The teacher spent class time showing students how the solution could be communicated in two ways.

A chemist has 2 identical test tubes able to hold 50 mL of liquid. In a beaker she has 48 mL of solution. She wishes to measure out exactly 42 mL of liquid. How should she proceed?

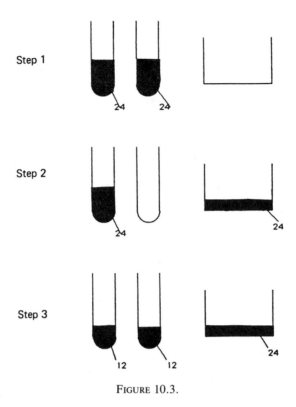

FIGURE 10.3.

After the teacher led the class to discover the first few steps in the process, he showed them how they might record their solution so that he could understand it. On the chalkboard, he recorded a student's description of the first few steps of the solution:

1. Divide 48 mL into two equal parts in the test tubes.
2. Dump 1 test tube back into the beaker.
3. Take the full test tube and put some back into the empty test tube so they are equal.

The teacher then showed how pictures could be drawn to represent each step (Fig. 10.3).

For this teacher, it was important that the student be able to articulate in some manner how a solution was reached. Grading reflected this value; for example, the single answer of "240" to the question, "How many squares on a checkerboard?" received 5 points, whereas a chart showing the pattern that led to that answer received 20 points. With this method, both the answer and the process received attention; particular emphasis was placed on the student's ability to communicate the thinking process he or she employed.

**Today's
Goal**

**Class Points**

_____ 1. **Work Period** _____ (5)
—on task
—
—
—

_____ 2. **Presentations** _____ (5)
—clear
—strategy used
—support for position
—

_____ 3. **Class Discussion** _____ (5)
—appropriate remarks
and questions
—

**Total** _____

*Mystery Person* _____

FIGURE 10.4. Problem-solving overhead transparency.

STRATEGY THREE

Another system for evaluating problem-solving performance and holding students accountable de-emphasized the answer. Students were evaluated on how they worked cooperatively to solve the problem, how they presented solutions to the class, and how the class discussed the problem solutions presented.

The teacher who used this system presented only one problem per lesson. She had a problem-solving overhead transparency that she used to structure the lesson (Fig. 10.4). She recorded the title of the problem on the transparency; for example, the wastebasket problem previously described was called *Wastebasket Whim*. She asked the class to decide the number of points they would work toward; students could earn up to a maximum of 17 points; 12 was the minimum target.

Students worked in cooperative groups that were randomly formed by having students count off by six or seven to form groups of four. Each group member took a letter (*T, E, A, M*). The teacher used cards labeled by letter and number (T-1, E-3, etc.) to randomly select students to present solutions toward the end of the class period and to randomly select a mystery person. This mystery person was observed throughout the lesson. If the mystery person was on task throughout the lesson, the class earned two extra points.

The procedure used was as follows: The problem was presented to the class. Clarification questions were answered. The teacher recorded the number of minutes the groups had to work on the problem. The teacher also recorded what she would be looking for during this work period. For the wastebasket problem, she recorded the following:

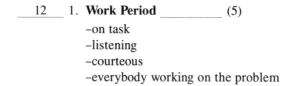

    12    1.  **Work Period**        (5)

       –on task
       –listening
       –courteous
       –everybody working on the problem

During another class, she might list other qualities, depending on the particular problem and the lesson objectives.

After the work period, the teacher gave students feedback on the number of points earned during that time span and reasons for losing points. She then gave them a few minutes to plan their presentation, the second phase of the problem-solving lesson. Because each presenter was randomly selected, groups used their planning time to make sure each member understood the solution process. On the overhead, the teacher wrote what she would be looking for during the presentations. For the wastebasket problem, she recorded the following:

          2.  **Presentation**        (5)

       –clear
       –strategy used
       –support for your position
       –drawings, math work used

During the presentation, class members were always polite, usually applauding at the conclusion of the presentation. Students were not penalized if they did not have an answer. They were, however, expected to explain the work they did in the time allotted. The number of presenters varied depending on the amount of time remaining after the work period. After the presentations, the teacher gave students feedback and recorded the number of points earned. For this problem, for example, the teacher commented that only one group explained the strategy that was used, so the class received 4 out of the 5 possible points.

During the presentations, questions could not be asked; this was reserved for the class discussion. During the class discussion time period, students could ask questions of other students or the teacher. They could not ask, "What is the correct answer?" Class ended with the teacher recording the number of points the class earned for the discussion and the number of points earned by the mystery person. Students tallied their total points to see if they had made their goal.

Depending on the activity and a teacher's particular objectives, each evaluation method is a viable method for evaluating students' problem-solving performance. Two of these evaluation systems emphasized the role of specific strategies. In all classes, students were observed using such strategies as building a table, guessing and checking, and drawing pictures. Specific strategies were either taught or referred to by all teachers. Teachers sometimes gave clues as to what strategy might be helpful or directly modeled how to use a particular strategy. In general, students seemed to have tools for solving problems.

## Cooperative Groups

The use of cooperative groups was another common characteristic of these classes. In some classes, students were randomly assigned to cooperative groups, and they were expected to work together; for others, a less formal approach was taken. Students either self selected who they would work with or decided to work alone.

During the interviews, teachers emphasized the importance of working in cooperative groups. Some of the teachers' comments include the following: "Group work helps deal with student frustration." "When they have others to share that feeling with, they keep working." "Gets too frustrating working alone." "It is more fun to work in groups." "Anxiety level is reduced." "They ask more questions when working in groups." "They act more independent." The teachers clearly felt that one important explanation for students' positive reaction to these lessons was the opportunity to work and talk with others about the problems.

## Regular Problem Solving

Another common element among these classes was that problem solving was done on a regular basis. In the interviews, all the teachers except one expressed that they were still not doing as much problem solving as they would have liked, considering that it was one of their most important goals. The teachers said that they were striving to do more problem solving and hoped to more fully integrate problem solving with the traditional curriculum. One of these teachers commented that, if he had the time, he could see teaching all mathematics through problem-solving lessons.

## Teacher as Problem Solver

All the teachers described themselves as problem solvers. They loved to solve problems, and they hoped their enthusiasm was transferred to their students. The interviews revealed that five of the six teachers had special training in teaching problem solving: four at the University of Minnesota in a Ford Foundation urban collaborative summer institute, and one teacher at the University of Texas at Austin. They believed their training made them better problem solvers— more open, for example, to using problems for which they did not initially know the answer. The teacher with the least experience was a recent graduate; she commented that her undergraduate training emphasized the teaching of problem solving.

## Summary

The classroom environments described in this chapter showed that the teachers involved in this study had orderly, business-like classrooms with a lack of discipline problems. The classes were *friendly* places, where the *students seemed to like the teacher*. The teachers had systematic methods for evaluating students' problem-solving performance and held students accountable for assigned tasks. They frequently used cooperative groups that allowed students to learn and find support from their fellow students. Problem solving was done on a regular basis, although still not as often as the teachers would have liked.

Three dimensions of the affective domain were prevalent in the classrooms we observed. Basically, there was evidence that students were willing to attempt problems, persevered in seeking a solution, and enjoyed problem-solving lessons. Lester (1980), in his review of the literature on problem solving, indicated that these three affective factors (willingness, perseverance, and self-confidence) are some of the most important influences on problem-solving performance. Reyes (1984) has also suggested that the students' willingness to work on a variety of mathematics tasks and their persistence in dealing with these tasks might make a difference in the degree to which a class is task oriented and easy to motivate. Thus, there were two common, yet distinct, themes in the classrooms of these exemplary teachers: the presence of highly prized affective outcomes and carefully organized classrooms in which the problem-solving teaching practices had important common elements. This suggests that developing positive affective responses to mathematical problem solving may involve classroom practices other than those specifically oriented toward providing a warm, caring classroom environment.

At this point, we are unable to draw conclusions about specific relationships between the structures and methods used by the teachers in their problem-solving lessons and the willingness, perseverance, and self-confidence demonstrated by their students. Based on our study of this select group of teachers, however, we do believe that there are important links between the way teachers organize and conduct problem-solving lessons and the affective responses and tendencies of their students. No doubt the relationships are complex. A particular classroom organizational scheme, such as small, cooperative groups, or a well-organized classroom routine is not likely, in isolation, to have a pronounced long-term impact on the cognitive or affective growth of students in the problem-solving domain. Rather, we must determine the means by which particular classroom structures and teaching behaviors interact with the way students function in learning to solve mathematical problems in classroom settings. Our research suggests that certain teaching conditions and structures may facilitate, or even be prerequisites for the development of positive affective responses to problem solving. These connections require and merit further research.

*Acknowledgments.* The research reported in this paper was supported by the National Science Foundation under Grant No. MDR-8460265. All opinions and recommendations expressed are those of the authors and do not necessarily reflect the position of the Foundation. We gratefully acknowledge technical research assistance from the Center for Research in Social Behavior at the University of Missouri and thank Cathy Luebbering for typing the manuscript.

## References

Good, T.L., & Biddle, B.J. (1988). Research and the improvement of mathematics instruction: The need for observational resources. In D.A. Grouws & T.J. Cooney (Eds.), *Effective mathematics and teaching* (pp. 114–142). Reston, VA: National Council of Teachers of Mathematics and Hillsdale, NJ: Lawrence Erlbaum Associates.

Grouws, D.A. (1985). The teacher and classroom instruction: Neglected themes in problem-solving research. In E.A. Silver (Ed.), *Teaching and learning mathematical problem solving: Multiple research perspectives* (pp. 267–280). Hillsdale, NJ: Lawrence Erlbaum Associates.

Lester, F.K. (1980). Research on mathematical problem solving. In R.J. Shumway (Ed.), *Research in mathematics education* (pp. 286–323). Reston, VA: National Council of Teachers of Mathematics.

McLeod, D.B. (1985). Affective issues in research on teaching mathematical problem solving. In E.A. Silver (Ed.), *Teaching and learning mathematical problem solving: Multiple research perspectives* (pp. 267–280). Hillsdale, NJ: Lawrence Erlbaum Associates.

Reyes, L.H. (1984). Affective variables and mathematics education. *The Elementary School Journal, 84*(5), 558–583.

# 11
# Affect and Problem Solving in an Elementary School Mathematics Classroom

ALBA G. THOMPSON and PATRICK W. THOMPSON

It is increasingly popular for research reports to acknowledge the importance of affect in mathematical problem solving. Unfortunately, merely acknowledging its importance does not get us far in understanding the role affect plays in problem solving. In the absence of efforts focused on understanding the role and importance of affect in problem solving, our attempts to deal effectively with students' affect during problem-solving instruction will be guided by hunches and good intentions.

Materials designed to help teachers with problem-solving instruction consistently include suggestions about the importance of attending to students' affect. Charles and Lester (1982), for example, stated the following:

It is absolutely essential that the classroom climate be conducive to problem solving! In fact, experience suggests that the classroom climate is so important in developing a successful problem-solving program that establishing an atmosphere conducive to problem solving should be the primary goal for about the first two months of the school year! The development of students' problem-solving abilities should take a back seat to the development of a positive atmosphere at the beginning of the year (p. 29).

Charles and Lester recommend that goals of a problem-solving program be set initially at developing willingness and perseverance in students, thus avoiding situations that might cause frustration. Recommendations to teachers that address the importance of developing students' confidence, willingness, and persistence are often included in instructional materials for elementary school students (e.g., Ohio Department of Education, 1980).

In his "Practical Suggestions for Developing Ability in Problem Solving," Brownell (1942) urged teachers to encourage students to "guess courageously." Davis (1987) emphasized the role of courage in "taking charge" in problem solving. Neither Davis nor Brownell, however, provided specific guidance to teachers on how to help students develop these qualities.

Cognitive researchers and developers of instructional materials acknowledge the importance of students' affect in developing problem-solving abilities. Neither group, however, offers specific instructional techniques that might be used to influence students' affect in a positive way. For the time being, this situation is natural, for not until we have a better grasp of the role of affect and its influence on problem-solving performance can we begin to formulate more specific instructional techniques. In the meantime, broad, general recommendations are the best we can offer.

As we offer broad, intuition-based recommendations, however, we should seek to learn more about how teachers interpret and make use of them in classroom instruction. How teachers interpret recommendations and how they translate their interpretations into practice are questions that should receive more attention from researchers than they have in the past. Teachers' insights into problems of dealing with students' affect can increase our understanding of the role of affect in problem solving and provide specific directions on how to deal with students' affect during instruction.

In summary, research has increased our awareness of the importance of affect in problem solving, but it has not increased our understanding of the role affect plays. Instructional materials offer general recommendations that deal with affective issues, yet they provide little specific guidance to teachers for their implementation. Also, we know little about what teachers make of recommendations that are given them, and we know even less about how they actually deal with students' affect during problem-solving instruction regardless of any specific recommendation.

## Relationship Between Students' Beliefs and Affect: Mandler's Discrepancy Theory

Any teacher who makes a serious commitment to the goal of developing students' problem-solving abilities will have to deal with the effect the new experiences are likely to have on the students' affective state. Because many recommendations for problem-solving instruction call for radical departures from what teachers and students expect to happen in the mathematics classroom, problem-solving instruction has great potential to create discrepancies and conflicts that are likely to cause strong affective responses in students (and teachers). Brownell (1942), for example, recommended that teachers "maintain a state of suspended judgment" on students' suggestions of possible solution strategies. His intention was for teachers to get students to evaluate and criticize each other's suggestions. It seems likely that a student who is "new" to problem solving and who views a teacher as a dispenser of knowledge would experience discrepancies that might result in strong emotional responses.

Just as students have well-formed beliefs and expectations about the roles of teachers, they also have beliefs about the nature of problem solving that will have to be modified if they are to benefit from problem-solving instruction. For example, a prevailing belief among students new to problem solving is that the objective of any problem-solving experience is to obtain a correct answer, that there is little value in having attempted a solution that does not lead to a correct answer, and that an answer is incorrect unless it matches "the" answer. Consequently, such students become quite frustrated when they fail to obtain a solution to a problem; they become even more frustrated when their solution is not "correct." In order to help students cope with frustration, their beliefs about the intended purpose of problem solving must be modified.

It is also common for students to believe that, unless one knows and remembers a direct path to the solution of a problem, there is nothing one can do to solve the problem. As a consequence, feelings of helplessness set in that often lead to high levels of anxiety. To overcome such helplessness, students must alter their original views and beliefs about the nature of problem solving.

Mandler's (this volume, Chapter 1) discrepancy theory provides an explanation for how students' beliefs and their interplay with problem-solving situations lead to affective responses. In Mandler's theory, an emotion is derived from an evaluation that something is not going as expected (a discrepancy) and a consequent response from the autonomic nervous system. According to Mandler's theory, when instruction in the classroom is radically different from what students expect, they will experience discrepancies between their expectations and their experience, and these discrepancies are likely to result in strong emotional responses. One might expect such discrepancies on a wide scale if recommendations for integrating problem solving into the curriculum are heeded.

Mandler's discrepancy theory suggests what may be a useful perspective for research on belief systems and affect. If, in fact, emotional reactions result from discrepancies between what is expected and what is actually experienced, then it should be possible to trace students' affective reactions back to the beliefs and expectations from which they arose. An understanding of the set of beliefs and expectations that students bring to the mathematics class would be a necessary first step in learning to deal effectively with their affect during problem-solving instruction.

## Affect in a Fifth-Grade Mathematics Classroom

The data presented here were obtained as part of a study of problem-solving instruction in an elementary school mathematics classroom. The purpose of the study was to examine the practice of a teacher who had participated in an in-service program on problem-solving instruction and who had taught problem solving on a regular basis as part of mathematics instruction. This chapter describes this teacher's instructional practice during problem-solving sessions. We attempt to document how he dealt with affective issues and the sources that influenced or informed his practice. This chapter also reports data on the affective states of three students in the teacher's class during problem-solving instruction.

### Subjects

The subjects in the study were a fifth-grade teacher, Mr. E, and three of his students. Mr. E taught mathematics, science, and social studies to two fifth-grade

|  | | Aptitude | |
|---|---|---|---|
|  | | High | Low |
| Attitude | Positive | Gail | Charles |
| | Negative | | Marcus |

FIGURE 11.1.

classes in a public elementary school (K–6) in a small city in central Illinois. Two criteria were used to select a teacher for this study. One was that the teacher be someone who was familiar with at least some of the current literature on problem-solving instruction. The other criterion was that the teacher be someone who taught problem solving regularly during mathematics instruction. Mr. E was identified by the district mathematics supervisor as satisfying both of these criteria.

The three students were selected on the basis of their scores on an instrument developed to measure students' perceptions of their own mathematical aptitude and attitudes. We sought students who would provide contrast in our analysis, thus, we identified one or two students (on the basis of their scores) for each of four cells in an aptitude-by-attitude matrix, with high and low as levels of aptitude and positive and negative as levels of attitude. To increase the reliability of our selection, Mr. E was asked to identify one or two students for each cell, independently of our assessment. For all but one cell we were able to select one student who had been identified by both methods. The three students thus selected were Gail, Charles, and Marcus (Fig. 11.1).

## Selection

The instrument used to select students consisted of released items from the attitudinal scales of the National Assessment of Educational Progress (mathematics component). The instrument addressed six aspects of students' aptitudes and attitudes:

1. Perceptions of the ease or difficulty and their like or dislike of the six major subjects of the school curriculum.
2. Feelings about specific aspects of doing mathematics, particularly problem solving.
3. Like or dislike of classroom activities associated with mathematics instruction, such as taking tests, listening to the teacher explain, discussing ideas in class.
4. Perceptions of the ease or difficulty and their like or dislike of various topics and processes in the mathematics curriculum, including problem solving.
5. Views about the nature of doing and learning mathematics.
6. Frequency of involvement in activities associated with successful performance in mathematics, such as helping classmates and working advanced problems alone.

## Procedure

Each student was observed and videotaped during one in-class problem-solving session. The student was interviewed following the class session in which he or she was videotaped. Each interview followed the same format. Each student was asked to explain his or her solution or attempted solution using the student's written class notes. The student was shown segments of the videotaped problem-solving lesson and was asked to talk about thoughts and feelings at different points of the lesson. The student was asked to comment on such things as facial expressions and other physical actions that might have been indicative of affective states.

The data from the teacher were obtained from classroom observations and interviews based, in part, on observed classroom events. All interviews were audiotaped. All the data were obtained in February, 1987.

## Results

The discussion of the findings is in three parts. The first part provides a general description of the climate that prevailed in Mr. E's class during problem solving and the organization of those sessions. The second part describes how Mr. E dealt with students' affect in problem solving. Finally, the data from each student are interpreted and summarized.

### THE CLASSROOM CLIMATE

Students' responses to the instrument were compatible with the generally positive atmosphere that prevailed in Mr. E's mathematics class. All but a few students indicated that they liked mathematics and were favorably disposed towards problem solving. Although overt signs of enthusiasm were rarely observed, a general impression of the students' attitude was, "We like it here."

Problem-solving sessions involving nonroutine problems were typically held once a week. Mr. E chose the problems from Addison-Wesley's *Problem-Solving Experiences in Mathematics, Grade 5* (Charles, Mason, & Gallegher, 1985). He followed the booklet's suggestions closely, rarely supplementing suggested questions with his own. The observed problem-solving sessions were conducted with students working in groups of four. In each case, a nonroutine problem was chosen, and approximately three fourths of the period was spent solving the problem and discussing solutions. Students worked on a written assignment, unrelated to the problem-solving lesson, during the remainder of the period.

Mr. E's first move in every observed lesson was to have students arrange their desks to form groups of four with two pairs of students facing each other. The class generally reacted favorably to the idea, "Today we'll do problem solving," and diligently proceeded to arrange their desks. Students were not assigned to specific groups. The groups were formed typically by students who sat near each other, although, on several occasions, some students joined others not in their vicinity. Mr. E would distribute a copy of the problem statement to each student and ask for a volunteer to read the problem aloud. Seldom did Mr. E want for volunteers.

After the day's problem was read, Mr. E asked questions about the problem statement in order to determine students' understanding of the situation, many times drawing attention to quantitative relationships described in the problem statement. His questions generally called for simple one- or two-word responses and were usually specific to the information given. Typically, his last question was, "What are we asked to find in this problem?"

Students appeared relaxed when questions were posed. At times, questions highlighted aspects of the problem situation that caused some students to become playful. Mr. E was tolerant and patient, maintaining a soft manner even when calling students back to task. It was common for one-third to one-half of the class to volunteer answers to Mr. E's questions.

Mr. E handled students' responses consistently. He would affirm a correct answer by repeating it. He would indicate that an incorrect answer did not seem to agree with the information given and would ask students to reconsider the problem statement. He would then restate the question and call on another student.

Following what was apparently the "understanding phase" of a solution process, Mr. E would call students' attention to what he referred to as "our list of strategies" — a list containing names of strategies (e.g., guess and check, and look for a pattern) that was displayed on a bulletin board in the back of the room and on a wall poster at the side of the room. The following excerpt is illustrative of how the phase "choose a strategy" was implemented. The problem was as follows:

Some elves were having a convention. The first time there was a knock at the door, 1 elf entered. On each of the following knocks, a group of elves entered that had 2 more elves in it than the group that entered on the previous knock. On the tenth knock, all of the elves had entered the convention hall. How many elves were at the convention?

| | |
|---|---|
| Mr. E: | Now, look at our strategies. It's either on the back board or up here [in the front of the room]. What strategy do you think you might try to use to solve the problem? Dana, what would you try? |
| Dana: | I'd say, draw a picture. |
| Mr. E: | You might draw a picture. Bryan? |
| Bryan: | Make a table or chart. |
| Mr. E: | Bryan might make a table or chart. Marsha? |
| Marsha: | Work backwards. |
| Mr. E: | Work backwards. Jason? |
| Jason: | Guess and check. |
| Mr. E: | Jason is going to guess and check. Kara, any different strategies? |
| Kara: | Make an organized list. |
| Mr. E: | Make an organized list. Charles? |
| Charles: | Find a pattern. |

Mr. E:    You might find a pattern. These are all strategies that might well work. Remember, work together as a group. You must all agree on the same solution. The solution must be presented in sentence form. You ready? Go right ahead and try your strategies. How many elves are at the convention?

Mr. E did not ask students to provide rationales for their suggestions. Although it was apparent that students were simply reading strategies from a list, Mr. E typically accepted all suggestions without evaluation. Relative merits of suggested strategies were never discussed before students tried their own.

While students worked in groups, Mr. E circulated among groups, listening to their discussions. Students stayed on task and interacted with one another. A leader was easily identifiable in most groups. Sometimes, while the students worked in groups, Mr. E would designate a person from each group to present their solution to the class.

When group work was done, Mr. E asked a student from each group to write the group's solution on the chalkboard. Mr. E asked questions about the various solutions presented at the board and summarized the groups' work. The solutions put on the board seldom differed in strategy—in direct contrast to the variety of solution strategies initially suggested at the beginning of the problem session. At the end of each observed session, Mr. E praised the groups for their "team efforts."

### DEALING WITH STUDENTS' AFFECT

Mr. E's concern for maintaining a positive environment in the classroom was an overriding one. His remarks during the interviews revealed that many of his actions were driven by some consideration of students' affect. In particular, he identified the following actions as ones he consciously chose to promote students' confidence in problem solving.

1. Have students work in groups, which takes the pressure off the individual.
2. Praise students for sharing ideas and suggestions.
3. Be encouraging.
4. Allow them time to think and work things out.

Mr. E acknowledged that there were differences among students in their motivation to solve problems. He indicated that, at the beginning of the year, the lack of motivation was general, and that most of the students were negatively disposed toward problem solving. He attributed students' negative dispositions to their lack of exposure to problem solving in previous years. He noted: "It's taken a lot of nurturing and prodding. . . . One of the things that I think has helped is picking problems that are interesting, problems that appeal to them . . . having strategies that they can use. . . . I try to take the pressure out of it, too, and working in small groups is one way to do that."

It was apparent from Mr. E's comments that he was a strong advocate of small-group problem solving. He reportedly learned about the use of small groups in a problem-solving course he had taken as part of a professional development program. When asked if he was bothered when a student relied entirely on other members of the group to solve a problem, he responded: "I suppose that can always happen, and it's hard to keep it from happening. But even then, I think they would learn from the others. Eventually something will stick. . . . (*Pause*) You know, you can't force kids to think. The best you can do is give them the opportunity, but you can't force them. I encourage them, and I try to make them feel proud of themselves when they do it."

When asked if he had ever observed signs of frustration in his students while they worked on a problem, Mr. E said that he saw frustration more frequently at the beginning of the year than now (February). He seemed to think that frustration was often associated with competitiveness, and that it was manifested in students who wanted to be the first to solve a problem and, therefore, were frustrated when someone else was first. As for the weaker students, he explained that "taking the pressure off getting the answer . . . letting them know what matters is the process" appeared to put them at ease.

We had become aware of a tendency on Mr. E's part to accept students' contributions without probing or asking them to justify or support their remarks. The issue was raised with regard to his acceptance of a variety of strategies suggested by the students, when it appeared that the students were offering strategies without a rationale. His response indicated that he saw more benefit in letting students attempt to use their own strategy and if inappropriate, discard it on their own, than in discarding it with his guidance. We also asked about his failure to probe students' understanding as they wrote their groups' solution on the board. He expressed a strong belief in "not putting students 'on the spot.'" He alluded to specific students' personality traits and the risk of embarrassing them and "turning them off."

Mr. E did not explicitly address students' beliefs as a source of conflict; yet it was implicit in his remarks that most of the affective difficulties experienced at the beginning of the year were mainly due to the students' view of problem solving as being limited to solving routine word problems and to their belief that the goal of problem-solving activities is to get "correct" answers. It was apparent that frequent exposure to problem solving in Mr. E's class had helped students broaden their view of problem solving and that, through his "nurturing and prodding," they had become generally comfortable with the activities conducted in class. Examples of this can be found in the following data on Marcus, Charles, and Gail.

## Marcus

Marcus' responses to the attitude instrument indicated that he perceived himself as having a low aptitude for mathematics and that his attitude toward problem solving was somewhat negative when compared to that of the majority of the students in the class. The problem given during Marcus' observation was the elves' problem. Marcus was visibly excited and eager as he joined his group. He sat with his leg underneath him, bouncing on it, and leaned over the desk to watch what one of the students wrote on paper. There was a lively discussion between two members of the group as they wrote down "1 3 5 7 . . ." and looked for a pattern. Marcus watched quietly, bouncing on his leg, leaning forward over his desk. The fourth student sat passively, watching and apparently listening.

Mr. E moved around the room; after a while he began designating a student from each group who would later put his or her group's solution on the board. Mr. E picked Marcus. Marcus looked at Mr. E with surprise and immediately began writing on his paper. Marcus looked frequently from his paper to the paper on which the other two students were recording their work. A few minutes went by before Mr. E called on students to come to the board. Strangely, when he came to Marcus' group, he called on a different student. (Mr. E later indicated that he had forgotten he had chosen Marcus.) Marcus leaned back on his chair in apparent relief and watched as the other students wrote on the board. He appeared attentive for most of what followed. At the end of the session, we asked Marcus to come with us to an adjacent room for the interview. He followed us playfully, looking back over his shoulder to the class, smiling and waving goodbye with his paper in his hand.

I asked Marcus to explain what he had done to solve the problem. He held the paper to me so that I could see what he had written, but he said nothing. Marcus had recorded the following on his paper:

$$1\ 3\ 5\ 7\ 9\ 11\ 13\ 15\ 17\ 19\ 21\ 23\ 25\ 27\ 29\ 31\ 33\ 35\ 37\ 39$$

I asked what "all those numbers" were. He read his list aloud. I pressed for a more thoughtful response. After some probing, it was obvious that Marcus did not know what the numbers he had recorded stood for. I asked him several times to read the problem to see if he could establish some connection between the story and the numbers, but this appeared to confuse him more. This is what followed (I = interviewer, M = Marcus):

I:     Why were you so excited when Mr. E gave the problem?
M:    I like this problem.
I:     Why do you like it?
M:    I like elves.
I:     Did you like solving the problem?
M:    Nope.
I:     How come?
M:    It was hard, really hard!
I:     Do you like to solve problems?
M:    Uh, not all the time. Sometimes. Sometimes I like to do it sometimes, not all the time.
I:     When do you like to do them?
M:    When it's easy.
I:     When is it easy for you?
M:    Like um . . ., the first time there is a knock at the door . . . and like . . . there was 5 elves and then 8 elves . . . 8 other elves came in. How many times . . . like times . . . I like times.

I:    Do you like it when you know what to do?

M:    Yeah!

I:    Do you get nervous when you don't know what to do?

M:    Sometimes.

I:    What do you do then?

M:    (shrugging his shoulders and pausing) I don't know.

I:    Do you ask for help?

M:    Uh-huh (yes). Sometimes. Sometimes I ask what we have to do, and I do it.

We viewed the segment of videotape that showed Mr. E choosing Marcus to record his group's solution at the board. I asked Marcus how he felt at that moment. Marcus shrugged his shoulders, smiling. He finally said; "I wrote it [the solution] down." I asked if he was glad that Mr. E did not send him to the board; Marcus nodded "yes."

Throughout the observations and the interview, Marcus showed no signs of uneasiness or distress. Given his inability to deal with problems, it was somewhat puzzling not to find such signs. It seemed that Marcus had not realized that there was something he was supposed to understand to be able to figure out a problem. Furthermore, it appeared that Marcus' idea of problem solving was to await instructions on how to do a task, and then to follow the instructions. These views, and the fact that the problem-solving sessions were conducted in a way that allowed Marcus to get assistance from his classmates, help explain, in part, his failure to experience uneasiness. There were few occasions when Marcus' idea of what was being expected of him clashed with any constraint of small-group instruction. Perhaps this would not have been the case if Mr. E had put Marcus in situations that required a display of understanding.

CHARLES

Charles' responses to the attitude instrument indicated that, although he generally enjoyed mathematics, he perceived himself as not being very strong in the subject, especially in problem solving. He had a lively personality and frequently volunteered in class. During the problem-solving session, he was actively involved in discussion with the other members of his group. He listened to others and contributed to the discussion.

In the interview, Charles articulately explained the solution that his group had produced. He followed his notes while explaining. His remarks revealed a good grasp of the problem and its solution. He was in control and appeared to enjoy it. There was no indication that he had any difficulty. His responses to the attitude instrument regarding his perceived ability to do mathematics appeared to be in sharp contrast with the observed behavior. We asked him what he did when a problem was difficult (I = interviewer, C = Charles):

I:    Do you always find problems this easy to solve?

C:    No way!

I:    What do you do when you get a problem you don't know how to solve?

C:    I think about it. Some problems are easy. . . . I can do those, but sometimes . . . it's like someone has a  . . . some people . . . you get advice when some people don't agree . . . and people who don't agree can help you work it out in a group.

I:    What do you do when there's nobody to help, like if you get a problem on a test.

C:    That's tough! But, sometimes I get it right.

I:    Do you ever get problems on a test?

C:    Yeah, sometimes.

I:    How do you feel when you get a problem on a test and it's hard?

C:    Well . . . I try to think of what to do.

I:    Do you get nervous?

C:    Yeah, sometimes. I usually get some of it right, though.

I:    You mean, even if you don't get the answer, you get some of it right?

C:    Yeah!

I:    How do you feel when you can't get the answer? Or you get an answer and you don't know if it's right?

C:    I hope that it's right.

I:    Do you ever wonder?

C:    Yeah.

I:    Do you try to find out if it's right or not?

C:    I check it.

I:    How?

C:    I take it and, like, go and put the answer down and see if it fits, times it or whatever.

I:    What if you find you've made a mistake?

C:    I go back and try to figure out where I messed up at.

I:    How do you feel when you know you're right?

C:    I feel grrrrrrrreat!

Although the observation and interview revealed no signs of strong feelings or emotion, Charles seemed to have a favorable disposition towards problem solving. Despite his responses on the questionnaire, which seemed to indicate that he did not think of himself as strong in mathematics, he appeared to be resourceful and well motivated. He seemed to derive satisfaction from being able to solve a problem and to enjoy working with classmates towards a solution. It was apparent from Charles' remarks that small groups provided a context in which he did not experience feelings of helplessness or anxiety.

GAIL

Gail's responses to the attitude questionnaire portrayed her as someone who did well in and enjoyed mathematics. She preferred solving problems over working computational exercises. She was one of only four students who noted that they enjoyed working problems alone and that they often worked on unassigned problems.

During the problem-solving session in which she was observed, Gail began writing even before Mr. E instructed students to start working in their groups. Gail worked independently while the other three students interacted among themselves. She stopped after a while and looked at another student's work, then joined in the interaction. It appeared that Gail had reached a solution before she joined the discussion, for she held her paper for others to see, pointing at parts of it as she talked.

Gail was one of the students who were asked to write their group's solution on the board. She wrote on the board while looking at her paper. Gail's solution was correct, but she did not explain it. Typically, Mr. E was the person who interpreted presented solutions.

Gail was reserved during her interview. She confirmed having solved the problem on her own before interacting with the others in her group. When asked to explain her solution, Gail simply showed me her paper. I asked specific questions about what she had written. Her responses were brief but accurate. It was clear she had understood and solved the problem correctly. We then viewed the portion of the video in which she started interacting with the others. I asked her what she had said. She explained a difficulty one of the other students had in "seeing the pattern."

Consistent with her questionnaire responses, Gail's remarks during the interview indicated that she enjoyed trying to solve nonroutine problems. We see in the following excerpt that difficult problems were interesting to Gail, and that she enjoyed being challenged (I = interviewer, G = Gail):

I:    Do you ever come across a problem that is too hard?
G:    Sometimes . . . not really.
I:    Do you ever get stuck?
G:    Yes, but I usually figure it out.
I:    Do you feel uneasy when you get stuck?
G:    Uh-uh. Not really.
I:    Can you remember a problem that was really hard for you?
G:    Yeah . . . one my dad gave me once. It was about marbles that were alike but one was different.
I:    Were you able to solve it?
G:    Yeah.
I:    Why did you think it was hard?
G:    I had to think a lot. You only had three chances to figure it out. I could do it in four.
I:    Was it frustrating?
G:    Yeah! But I finally figured it out.
I:    How did you feel when you figured it out?
G:    Smart.

I:    Do you always feel smart when you solve a problem?

G:    Uh-huh. I guess that's what I like about it.

When asked whether she would rather work on a problem alone or in a group, she had no preference. She conceded, however, that she typically worked the problem on her own even when she was in a group. Confidence in her ability to reason her way to a solution and a sense of pride and satisfaction in being able to do so were reflected in her remarks as well as in her behavior.

## Discussion

The interviews with Marcus and Charles suggest that the use of small groups served to ease some of the pressure that students would have faced if they were to have solved their problems independently. The ability to rely on help from other group members minimized the number of occasions that students needed to struggle, alone, with a problem. Consequently, small groups helped reduce the likelihood that students would experience strong feelings of helplessness or anxiety and contributed to their generally favorable attitudes toward problem-solving experiences. This was consistent with Mr. E's objective in using small groups.

Small groups were not the only aspect of instruction to which the relaxed climate that prevailed in the class can be attributed. Mr. E's friendly and supportive manner and his tendency to accept students' contributions unquestioningly appeared to contribute to students' openness and to the low-stress atmosphere that prevailed in the class during problem-solving sessions.

We remain curious about the effects, if any, of asking students to support and evaluate their suggestions and solutions and of having them work independently. This is not to imply that there has to be a tradeoff between a classroom climate that is cognitively demanding and one that has a positive affect. Students can be successfully encouraged to participate willingly in low-stress problem-solving activities, as evidenced by Marcus and Charles; however, it is also possible to help students see that struggling toward a solution can be an intrinsically rewarding intellectual activity, as evidenced in Gail's case. Likewise, it should be feasible to develop in students an appreciation for the merits of evaluating and critically reflecting on their own and other's ideas. Developing an appreciation for "struggle" would have to be gradual, requiring a long period of time; however, it is unlikely to take place unless it is an explicit goal of instruction that is appropriately and frequently translated into classroom action.

The affective goals of problem-solving instruction should not be limited to getting students to enjoy problem solving. Positive affect in problem solving also involves an appreciation of intellectual activity. The absence of negative emotions, such as frustration and anxiety, among students need not be taken as an indication of a desirable affective state, because expert problem solvers also experience such emotions. The important issue is not whether such negative emotions occur, but how students cope with them *when* they occur and what drives students to persevere at a task even when it entails some degree of struggle.

A constraint under which Mr. E behaved was to avoid any situation with the potential for public failure. This led him not to press students for explanations, apparently for fear that incurring an error publicly might diminish students' self-esteem. This particular concern appears to be common among American educators.

In a recent report of a cross-cultural study of mathematics teaching and learning, Stigler and Perry (in press) document sharp differences in the way errors are viewed in Japanese, Chinese, and American schools. In Japan and China, for example, errors are not viewed as measures of intelligence; rather, they are viewed as an indication that one is still learning the subject matter. In Japanese and Chinese schools, success is not taken as an indication of ability or "smartness," which more often is the case in American schools; rather, success is viewed as the product of hard work.

Stigler and Perry relate an anecdote that illustrates the effects of an "objective" view of errors. While visiting a mathematics class in a Japanese elementary school in which the lesson concerned drawing cubes in three-dimensional perspective, they observed the following situation.

Each student was working in his or her notebook, but there also was a great deal of discussion from desk to desk, and the noise level was rather high. The discussions were not inappropriate, however; rather, it was directed almost completely to the mathematical topic at hand.

Against this background, one child was having trouble. His cube looked crooked, no matter how carefully he tried to copy the lines from the teacher's example. And so the teacher asked this child to go to the blackboard and draw his cube. Standing there, in front of the class, he labored to draw a cube correctly while the rest of the students in the class continued working at their desks. After working for 5 or 10 minutes, he asked the teacher to look at his product. The teacher turned to the class and asked, "Is this correct?" The child's classmates shook their heads and said, "No, not really." After some open discussion of where the problem might lie, the child was told to continue working at the blackboard and try again.

This scene continued for the duration of the 40-minute class. As the lesson progressed, the group of American observers began to feel more and more uncomfortable and anxious on behalf of the child at the board. We thought that any minute he might burst into tears, and we wondered at what he must be feeling. Yet he did not cry, and in fact did not seem at all disturbed by his plight. At the end of the class he finally drew a passable cube, in response to which the class applauded (Stigler & Perry, in press).

The preceding observations stand in stark contrast to our observations in Mr. E's classroom, and they stand in contrast to our knowledge of American classrooms at large. The essential distinction between the setting described in Stigler and Perry's observation and the setting in Mr. E's class appears to be the difference in the role of problem solving. The perspective of the teacher in Stigler and Perry's passage appears to be that problem solving is an activity necessary for the construction of knowledge. This was not the role given to problem solving in Mr. E's class, in which problem solving was a subject matter.

Stigler and Perry (in press) point out that, by looking only at classrooms from within a limited frame of reference, we "become accustomed to many of the most predominant characteristics of those classrooms, and thus fail to note the significance of those characteristics." Perhaps we need to re-examine the meanings we ascribe to success and to student affect and question the practices that we tend to take for granted as necessary for ensuring students' success.

## References

Brownell, W.A. (1942). Practical suggestions for developing ability in problem solving. In N.B. Henry (Ed.), *The forty-first yearbook of the National Society for the Study of Education* (pp. 433–440). Chicago: The National Society for the Study of Education.

Charles, R.I., & Lester, F.K., Jr. (1982). *Teaching problem solving: What, why & how.* Palo Alto, CA: Dale Seymour Publications.

Charles, R.I., Mason, R.P., & Gallegher, G.R. (1985). *Problem-solving experiences in mathematics, grade 5.* Menlo Park, CA: Addison-Wesley.

Davis, R.B. (1987). "Taking charge" as an ingredient in effective problem solving in mathematics. *The Journal of Mathematical Behavior, 6,* 341–351.

Ohio Department of Education. (1980). *Problem solving: A basic mathematics goal.* Columbus, OH: Author.

Stigler, J., & Perry, M. (in press). Mathematics learning in Japanese, Chinese, and American classrooms. In G. Saxe & M. Gearhart (Eds.), *Children's mathematics.* San Francisco: Jossey-Bass.

# 12
# Affective Factors and Computational Estimation Ability

JUDITH THREADGILL SOWDER

The purpose of this study was to examine the influence of affective factors on the computational estimation ability of preservice teachers. Past research has delineated the cognitive processes used by good estimators (Reys, Rybolt, Bestgen, & Wyatt, 1980) and poor estimators (Threadgill-Sowder, 1984). The good estimators in the Reys et al. study seemed to have acquired estimation skills without any formal instruction. Are there also noncognitive factors that influence people to acquire such skills?

Computational estimation skill, particularly when acquired without formal instruction and, therefore, free of imposed algorithmic structures, is closely akin to problem solving. This skill is usually a means to an end, rather than the end itself, and, therefore, can be selected or ignored while achieving a problem solution. Because one is rarely required to estimate and, therefore, does not do so unless comfortable with the process, it is relatively free of the emotion described by Mandler (this volume, Chapter 1) as the hot interpretation of affect. But it seems highly likely that there are cooler factors that influence whether or not an otherwise cognitively capable individual pursues a mathematical skill for which it is likely that no instruction has been provided.

Some of these cooler affective factors have been identified in previous research. Indeed, estimation ability and self-concept as a mathematics student have been found to be positively correlated (Bestgen, Reys, Rybolt, & Wyatt, 1980). This is not surprising, because mathematics self-concept would seem to influence the acquisition of mathematical skills acquired without formal instruction. Reyes (1984) noted that "confident students tend to learn more, feel better about themselves, and be more interested in pursuing mathematical ideas than students who lack confidence" (p. 560). Reyes noted a number of studies in which a significant positive relationship between self-concept and mathematics achievement has been found.

Whether a student attributes success and failure to internal or external causes, and whether such causes are under the student's control (Weiner, 1972), has recently been shown to be related to achievement in mathematics (Fennema & Peterson, 1984; Reyes, 1984). This factor is also related to mathematics self-concept. A student confident of his or her mathematical ability is more likely to attribute success to ability and failure to some other cause, such as task difficulty or lack of effort. A student lacking confidence is not likely to venture beyond the security of known algorithmic processes that are certain to lead to success. This student is not likely to be good at estimation and not likely to attribute success to ability.

A factor unique to computational estimation, but that seems affective in nature, is tolerance for error. Persons uncomfortable with estimates, rather than exact answers, are unlikely to estimate when they are not required to do so. This factor has been conjectured as an important influence on whether or not one becomes a good estimator (Reys et al., 1980; Rubenstein, 1985). It is a rather elusive construct, however, and one that is difficult to access. Too large a tolerance for error, such as when a student's misplacement of a decimal point makes an astronomical difference in an answer but is found acceptable by the student, is obviously undesirable.

The value accorded to estimation is also a factor worth considering. Good estimators say that estimation is an important skill and that they use estimation skills frequently in everyday life (Reys et al., 1980). This relation between perceived usefulness and acquired skill should also be true for mental computation, an important component of estimation. Attitudes toward the role of calculator usage might also help reveal an individual's comfort level with internal algorithmic processes. It is likely that other beliefs also influence the acquisition of computational estimation skills. For instance, if someone believes that there is always just one right answer to a problem in mathematics, then that person would be hard pressed to find that one correct estimate to a problem.

## The Study

All of the factors identified in the preceding section were considered in this study, which used written evaluation instruments and interviews. Fifty-eight students completed all of the written assessments of skills and affective factors. From this group of students, 17 were selected for interviews. The university-level students participating in this study were in the first of a sequence of three mathematics courses offered for prospective elementary-school teachers.

### Instruments

Three tests assessing estimation skills, attitudes, and attributions of success were developed. The two-part estimation test consisted of basic computation items with whole numbers, decimal numbers, and fractional numbers. Each of the first 14 items was paired with a target number. Students were required to tell whether the answer to the computation problem would be more than or less than the target number or whether they were unable to determine the relationship in the time allotted. For the last 10 problems, students were to give an estimate of the answer to a computation problem.

Using several sources, an attitude inventory of 70 items related to the factors of interest in this study was written. Items on *self-concept*, *enjoyment*, and *usefulness of math* from the Fennema-Sherman Mathematics Attitude Scales (Fennema & Sherman, 1976) were used. An example of such an item is, "I know I can do well in mathematics." In addition, a few items on effort as a mediator of mathematical ability, thoughts about success and failure in mathematics, and failure as an acceptable phase of learning mathematics were adapted from Kloosterman (1986). Finally, items composed by the experimenter that dealt with tolerance of error, common beliefs about mathematics, and uses of estimation, mental computation, and calculators were included. For each item, subjects were required to choose one of five responses: strongly agree, agree, undecided, disagree, strongly disagree.

The third measure used was an abbreviated version of the Fennema-Peterson Attribution Scale (Fennema & Peterson, 1984). The 24-item scale measured a student's inclination to attribute success to ability, effort, outside help, or task difficulty (three items each), and failure to each of the same four attributes (three items each).

## Procedure

The estimation test was administered by placing each item on an overhead projector for 10 seconds. Students had narrow strips of paper with the multiple choice answers (more than $x$; less than $x$; I don't know) for each of the first 14 items, and blank spaces for estimates of the last 10 items. There was little additional space available for scratch work, and students were not allowed to have other paper on their desks. At the completion of the estimation tests, students were given the attitude inventory and the attribution scale.

Fifty-eight students completed all three measures. Predetermined acceptable ranges were used for the last 10 items of the estimation test. Students who received very high scores (20 or more correct out of 24) and very low scores (10 or less correct) on the estimation test were identified. All 7 high-scoring students, and 10 of the 12 low-scoring students completed 45 to 60-minute interviews within 2 weeks of the in-class portion of the study.

Each interview had two distinct parts. The student was first asked to rework 16 of the class-administered estimation tasks and explain the process used to complete each task. The student was then shown a profile based on his or her responses to the class-administered estimation test and the other two survey scales. The profile consisted of my written notes on (a) estimation ability, (b) mathematics self-concept, (c) attributions of success and failure and whether the student thought about reasons for success and failure, (d) tolerance for error, (e) use of estimation, mental computation, and calculators, and (f) beliefs about mathematics. We discussed the accuracy of my notes on each item in the profile. In the interviews, students were encouraged to say whatever they wanted about the profile statements, their own feelings, attitudes, and backgrounds in mathematics, in general, and about estimation, in particular.

The estimation tasks from the interviews were then scored. Each of the 16 responses was assigned three points if the student used an efficient estimation process and arrived at a reasonable estimate. Two points were assigned if the process was generally correct but some error was made in mental computations or an inefficient process was used. For example, students were asked to say whether $283 \times 7$ would be more than or less than 1400. A student who said 200 $\times$ 7 was 1400, so $283 \times 7$ is greater than 1400 was assigned 3 points, whereas a student who estimated $280 \times 7$ is about 1900, again greater than 1400, was assigned 2 points. If a student did *some* estimating or indicated some insight into the estimation process but could not carry it through, 1 point was assigned. Students who attempted to mentally carry out the paper-and-pencil algorithms associated with the problems, could not do the problem, or made serious place-value errors, received 0 points. Using this grading scale, students fell more naturally into three groups rather than two: Group 1 (high, 36–47 points), Group 2 (medium, 20–32 points), and Group 3 (low, 5–15 points). The analysis to follow discusses students in terms of these categories. Group 1 had five students, Group 2 had seven students, and Group 3 had five students. In most instances, only the high and low groups are contrasted. When responses from the middle group do not fall along the expected continuum, or when they give additional information helpful to understanding results, they are included.

## Results

Because most of the discussion during the interviews followed the outline provided by the student-profile form, this form provides an organizational tool around which to consider differences between groups and between individuals in reporting the results. The final section of the results gives data from the interviews on students' perceptions of how they became good estimators.

### Ability to Estimate

As stated previously, the 17 students interviewed fell into three rather distinct categories when scores on estimation tasks were compared. Typical responses from the three groups on two estimation items should clarify the differences between the groups in terms of the grasp of the meaning of the numbers (number sense), the understanding of the operations used, the tendency to curtail thought when working through a problem, and the relative ease in carrying out a process once decided upon.

Problem: $55.65 - 26.95$
Choices:
  Greater than 20?
  Less than 20?
  I don't know.

Group 1 responses:

"55 − 25 is 30, so greater than 20."

"Add this [20] to that [26.95], I got 46." (The student selected *greater than 20.*)

Group 2 responses:

"56 − 27. About 60 times [*sic*] 27 would be in the 40 somethings. So it's going to be larger than 20."

"These numbers don't subtract easily from these ones, you know? But I could make it 31, but then that's not right. I could just say its 30. (Where did you get 31?) Because I was going to make this 25 instead of 27 so it'd be as easy as possible. So it'd be like 28, larger than 20."

Group 3 responses:

"Smaller than 20. Take 5 away from this [the 6 in 26], it'd be 21. It'd be a little over 20."

"Put 26.95 under the 55.65. Zero, crossed out the 5 . . . ."

Problem: $7/8 \times 21/31$

Choices:

Greater than 1?

Less than 1?

I don't know.

Group 1 responses:

"7/8, almost 1, so about 2/3."

"When you have two small fractions and you multiply them together, it's even smaller, so I say smaller than one."

Group 2 responses:

"$7 \times 20$ would be 140, $8 \times 30$ would be 240, so that would be smaller than one."

"I would say smaller than one because the ratio is going to be similar because the numerator is smaller than the denominator [7/8], and the same is here [21/31]. So the increase is not going to be such that the numerator will be greater than the denominator."

Group 3 responses:

"I would say larger because I'd get a big number when I multiplied them out."

"I'd probably just multiply straight across, like $7 \times 21$ and $8 \times 31$. Yes. Because I'd look across to see if 7 could go into 31, and it doesn't, so I just multiply across." (Gave *greater than one* as her answer.)

## Self-Concept in Mathematics

Of the five students in Group 1, three had very strong positive self-concepts regarding their mathematical ability, their self-assurance while undertaking mathematical tasks, and their confidence that they could and would do well in mathematics. The remaining two were less confident. Both had attempted higher level mathematics courses and done poorly in them, contrary to their initial expectations, which, in fact, probably lowered their self-concepts. (See later section on Stella.)

In contrast, the self-concepts of four of the five in Group 3 were much lower. Comments such as, "I've never been good in math" and "I get through it, but it's not one of my best subjects," were typical for this group. The one student who exhibited some self-confidence seemed to enjoy mathematics, at least she enjoyed paper-and-pencil algorithmic tasks, and had always experienced some measure of success. At one point during high school, she thought she wanted to major in mathematics, but then she encountered difficulty in intermediate algebra and lost interest. One might wonder if her confidence is related to the fact that she seems never to have experienced any math or test anxiety: "I don't know how I would feel to get tense. Like, everybody tells me that they get so scared when a test comes up. And I don't know what they mean when they say that they're scared."

Students had fairly accurate appraisals of themselves, if class work and grades are considered to be reflections of their ability. Four students in Group 1 received As in the course and the fifth (Stella), a B. The students in Group 2 received Bs and Cs, and students in Group 3 received Cs, a D, and an F for their final grades. It should be noted, however, that students in Group 3, and to some extent students in Group 2, believed that if they just had enough *time*, they could learn the mathematics they needed to know. Some students expressed the belief that they could do better in other mathematics classes. The student receiving a D, for example, often mentioned that he was doing well in an intermediate algebra class, so that it didn't make sense to be having difficulties in this "easier" course. Another (Group 2) student also expressed frustration with the course, claiming to get higher grades in other mathematics courses: "This class is so weird because I get so frustrated because I don't want to understand the concepts. I just want to do the problems. And I didn't realize that's what the class was [understanding concepts] until a little while ago."

## Attribution of Success

Students in Group 1 attributed their successes in mathematics primarily to ability, secondarily to effort. One student (Stella) distinguished between school mathematics, which she found somewhat difficult and in which she attributed success to effort, and daily-life mathematics, with which she felt very comfortable and in which success was due to her natural ability. Students in Group 3, on the other hand, attributed success primarily to effort and not at all to ability. This point, obvious from the results of the attribution scale, was strongly reaffirmed in discussions.

Few students in any of the groups claimed to reflect on their successes in mathematics, when such occurred. Some indicated that they saw no value in thinking about it, but it seemed simply not to have occurred to most students to do so.

## Attribution of Failure

Two of the five students in Group 1 claimed never to have experienced failure in mathematics and so were unable to select a factor to which they could attribute failure. The other three claimed that when they didn't do well, they either hadn't studied enough or the task was too difficult, particularly in the case of the two students who had not done well in the higher-level mathematics courses that they had taken. One student (Kathy, discussed later in this chapter), who had sometimes experienced difficulty with mathematics, placed some of the blame on the teachers: "When I got into the more complicated ones [courses], I didn't feel really good. I thought it was because [of me]. Until I sat down and thought, no, it's how well I understand, and if I explain what I don't understand to the teacher, and the teacher can't explain it back to me, that's the *teacher's* problem, being forced on me." Students in Group 3 attributed failure to lack of effort or task difficulty but, surprisingly, considering their low self-concepts in mathematics, not to lack of ability.

All students claimed to think about reasons for not succeeding. This was generally interpreted as not succeeding on a particular problem and/or not getting a high score on an exam. Typical reasons given were carelessness, not studying enough, and not understanding.

## Use of Mental Arithmetic

Students from Group 1 claimed that mental computation was easy for them and that they did not avoid it, whereas students from Group 3 either found it difficult or were undecided and were, for the most part, uncertain about whether they avoided it. Group 3 students seemed not to be aware of whether they did any mental computation. From observations of these students both during the estimation portion of the interview and their in-class work, they seemed to do little, if any, mental computation, except occasionally for trivial problems involving few digits, and no regrouping, such as $7 \times 21$.

## Use of Estimation

Four items were included in this category: (a) When I work a math problem, I check to see if the answer makes sense; (b) When I use mathematics in my daily life, I often use estimates rather than finding exact answers; (c) People use a lot of math in daily life without realizing that it is math; and (d) I estimate answers to arithmetic problems as a way of checking on my calculations. There were noticeable differences in responses between Groups 1 and 3 on the first two items, but not on the third and fourth items. Group 1 students strongly agreed with the statement about checking to see if an answer made sense, whereas Group 3 students either claimed to do so (agree, but not strongly agree) or were undecided on the item. In using estimates rather than exact answers in daily life, Group 1 students tended to agree or strongly agree, and Group 3 students only agreed or were undecided. Strangely enough, five of the seven students in Group 2 claimed *not* to use estimates in daily life. (One wonders, then, whether Group 3 students used estimation other than in a very gross fashion.) All students felt that people use a lot of math in daily life without realizing that it is math, and answers were mixed in all groups on the item regarding estimating to check answers. Several students said they preferred to use a calculator to check their work.

## Use of Calculators

Responses to calculator items proved interesting. Group 1 students disagreed with the statement that it was always better to use a calculator for calculations if one was available, but Group 3 students were uncertain. Only one Group 1 student preferred to do calculations on a calculator, saying that it was simply a time saver, whereas Group 3 students were again mostly undecided on this item. Group 3 students generally trusted calculator answers more that their own calculations, but this time Group 1 students were undecided. Several students observed that they did not know *how* to use a calculator other than for performing the four fundamental operations on whole or decimal numbers.

## Tolerance for Error

The several items included in this category were attempts to pinpoint the elusive phenomenon of tolerance of error, believed by some to be one of the hallmarks of a good estimator (Reys et al., 1980). Responses to the first item (It makes me uncomfortable to be wrong when a teacher calls on me in a math class) ranged from undecided to strongly agree by students from Group 3, whereas students in Group 1 disagreed, for the most part. Students in Group 1 felt that they would not *be* wrong unless the question was on new material or was particularly difficult, in which case it was not shameful to be incorrect. Students in Group 3 used this same argument to qualify their answers. They would not be embarrassed in such a situation, but they would be embarrassed if the question was one they obviously should be able to answer.

Other items were even less discriminating. On the second item (When I shop and the clerk makes a mistake that is more than a few cents off, I would notice the amount I am being charged is wrong), there was general agreement in all groups. Interview answers seemed to depend more on the age (and shopping experience) of students than on any other factor. There was also general agreement with the third item (When I do math problems, I feel better having the exact answer rather than a ballpark estimate). In each of the groups, responses to a fourth item (When I balance a checkbook and I'm just a little bit off, I let it go rather than spend a long time trying to locate my errors) ranged over the spectrum of choices. Students in Group 1 strongly agreed or agreed with the fifth item (I check my calculations at least once to make sure I'm right). Students in Group 3 gave responses in all categories. When questioned, they said that much of the time, particularly on tests, where it really mattered, they usually did not have time to check their work.

## Thoughts and Beliefs About Mathematics and Doing Mathematics

Group 1 students strongly disagreed with the statement that mathematics problems always have just one right answer. In Group 3, some students disagreed and some were undecided. Responses to the item that working hard increases ability in mathematics were just the reverse: Group 1 students disagreed or were uncertain; Group 3 students were generally in strong agreement with this statement.

## Becoming a Good Estimator

Good estimators were also asked to try to think about and attempt to explain *why* they had become good estimators. In each case, the student was able to give personal information that was thought to have influenced his or her facility at estimation. Such information is particularly useful because not all mathematically able students become good estimators. Among the original 58 students tested for this study, for example, 11 students appeared to be quite capable mathematically, at least insofar as they received As as final grades in the course, which emphasized conceptual understanding of number systems and which naturally required computational ability. Yet only four of them were among the initial group of good estimators, as determined by the estimation test given in class.

All five of the students from Group 1 were able to recount something in their past that stimulated them to become proficient in estimation. Two students spoke at length about their reasons for becoming good estimators. Both students were women in their early thirties, with families, who had returned to school after long absences, but the similarity seems to end there. The reflections of these two students (Stella and Kathy) provide insight into the different and unusual paths that can lead to becoming a good estimator and, therefore, are provided in more detail than the reflections of the first three students described here.

### MICKEY

Mickey, whose father had been a mathematics teacher, had decided to enter teaching after working several years as a carpenter. "We were always playing math games . . . . We would play card games that were math related. Even when I was out of school, I would find an old math book and just do equations. It's like doing puzzles. I think math is fun." Mickey wondered if his several years in carpentry had helped him develop a good number sense. He was careful to point out, however, that his background did not necessarily cause him to become a good estimator: "There were five kids in my family, and I'm the only one that grew up with number sense . . . . My two brothers are carpenters."

### MEG

Meg also grew up in an environment that encouraged the study of mathematics. "My mom introduced me to math, like, as puzzles. She had me doing algebra before the school did, just because she'd go 'Look! It's like a puzzle.' Then that would make me think of math as fun . . . . I got off to a good start." Meg claimed to use a lot of estimation in her daily life, particularly in her athletic training program.

## ROBERTA

Roberta was the only student who had received formal instruction in estimation. "I had a chemistry class with a really great teacher in high school. And estimation was extremely important to him. So a lot of rules of estimation were talked about. We even had estimation quizzes. [But] to some extent I would have already had it. I think the chemistry teacher just helped me get over some of the stupid mistakes [and helped me learn] some of the things which made estimation quicker."

## STELLA

Of the five students in Group 1, Stella had by far the most difficulty learning mathematics, although she appeared to be mathematically capable. She made a sharp distinction between everyday mathematics and classroom mathematics: "In terms of what I do day to day, I have absolutely no apprehension at all. Or even stuff that's more complex. Figuring out ratios and things like that in my head, I have absolutely no problem at all. It's when I'm sitting in a classroom [that I have problems]." During the second part of the interview, Stella was asked to react to written statements on her profile (she is unsure of herself while doing math; math is her worst subject and is difficult for her) that were taken from information provided in the attitude inventory completed earlier. (I = interviewer, S = Stella.)

> S: Well, you had it exactly right that I'm not comfortable *doing* math. I'm uncomfortable *with* math. I don't feel intimidated by that. I feel intimidated by *problems*, or that I'm getting a grade for it.
> I: In other words, you have no fear of using math in your own life?
> S: No, none at all.
> I: It's just what happens in class?
> S: Yes.

What is particularly interesting in Stella's case is the role estimation plays in her daily life, where mathematics is easy, and how she uses estimation to help her cope with school mathematics.

> I: I'd be interested to know how you came to estimate as well as you do.
> S: Well, I don't know. Except that maybe I use it more than others. Maybe because of what we were talking about earlier. What I had to do in a lot of my math was, instead of being comfortable thinking I had the right answers, what I usually had to do ahead of time was say, "Now what is my answer going to be close to?" Then I'd work at it and say, "You're way off." . . . And so maybe it [skill at estimation] was a

> defense mechanism that I developed. Because that's what I rely on more than my math.
>
> I:    Do you feel like you have a good sense of what a number means before you look at it?
>
> S:    I feel very comfortable with numbers. I don't feel comfortable doing math problems. I feel very comfortable relying on myself mathematically. But I don't feel comfortable doing algorithms, let's say.

Later, she discussed her attribution of success to effort.

> I:    But then, why are you successful in daily life? That's not due purely to effort, is it?
>
> S:    Oh no. In fact, to me, it's not effort at all. I don't have to go, "Oh, gosh, I'm going to estimate." It just flashes in my head, and I don't even think about it. It's easy for me, and it's not a struggle. So I guess you would say ability, too. That's why I'm making this distinction. Because it's not an effort for me to do math outside the classroom.

Toward the end of the interview, her feelings were summarized in this statement:

> S:    I don't think that I'm ever going to be comfortable with it. I'm never going to be fluent with mathematical problems. But I also do not feel that I am a mathematical cripple, because I function quite well and quite comfortably outside the classroom.

The distinction Stella makes between school mathematics and out-of-school mathematics is not an uncommon one. Maier (1977) claimed that whereas school mathematics is largely dependent on paper and pencil, "folk mathematicians rely more on mental computations and estimations and on algorithms that lend themselves to mental use" (p. 86). Lave (in press), describing the increasing body of research on the everyday mathematics of "just plain folks," noted the efficacy of the heuristic strategies these people use in dealing with mathematical situations in their lives, even though they often believe they are unable to do "real math."

Stella's discomfort with classroom mathematics may have stemmed from an experience with a probability course, which she found to be incomprehensible. She was not experiencing any real difficulty in the mathematics course for prospective elementary-school teachers, although she needed to spend more time on the course than other students in Group 1.

What are the ramifications of such a sharp distinction between school and everyday mathematics? One fallout was a strong negative feeling towards the use of calculators, which, she said, "makes you lazy." She does not allow her 10-year-old son to use a calculator. Whether she will change her mind about this issue by the time she begins teaching seems doubtful. Which of her two kinds of mathematics—classroom mathematics and everyday mathematics—will she teach to her students? Or will she attempt to help her students find such a distinction unnecessary?

KATHY

Kathy also learned to estimate as a means of compensating, but for different reasons than Stella (I = interviewer, K = Kathy):

I:    You have some very good estimation techniques. Where do you think you got them?

K:    I've had to start going for speed, and so I've had to streamline. I tend to get bogged down in detail, but since I decided that estimation is only a target, it's not the exact answer, I didn't have to be so concerned. And that took a lot of pressure off.

I:    How did you arrive at that? The conclusion that you don't always have to be exact.

K:    I finally realized that it's not making any difference. I'm putting on too much pressure. I don't have to be perfect. . . . It basically comes from letting go of having to be exact and having to push for speed, whereas before I would have preferred to work it out.

I:    This is something that has evolved for you recently then?

K:    This is recent. Just being able to let go . . . and say I can't have absolute accuracy and get through all of this.

I:    When you say "all of this," what do you mean? What have you been doing that has forced you to this realization?

K:    OK, let's see. My husband decided to go back to school and still work part time, and so I had apartment management. I had a child with learning disabilities that the insurance decided they weren't going to pay for, so I would have to do the therapy myself. And then involved in this school work, and then trying to hold onto my own life. It was the pressure.

I:    When you say you can let go, you don't have to be so stressed, this sounds to me as if this is a decision you've made that permeates much of your life. Does the estimation . . . . (interrupted).

K:    It became a part of it. Mathematics is very precise, and then estimation is the exact opposite. You don't have to be precise.

I:    And that became incorporated into your life?

K:    Yes, my whole attitude of thinking.

I:    Were you aware before this conversation that estimating had become a part of that change?

K:    I began to get the idea, especially when I noticed I was having to work for speed and basically when I started working with my [learning disabled] fourth grader this year. I realized that *it really doesn't matter*.

Kathy had also experienced difficulties with mathematics in her past. She did not do well in high school geometry and had apprehensions about taking and teaching geometry. Probably because she tends to be a doer, with a strong self-concept about her ability, she took steps to prepare herself for these events.

K:     I went to the library and got a book on studying, and hit the part on geometry, which always gave me fits, and found out basic errors I've made. I said, "Oh, that's why I didn't do so good." And so I figured when I took it again, I would have the tools I need to do well.

But even this strong self-concept was something that had evolved in recent years.

K:     Some things I plain backed away from, and I decided it was part of the way I was raised. The men were good in math and the women weren't. And I faced a conflict. Do I conform with these expectations, or what my abilities really are, and for a part of the time I conformed to the society. And I've been doing a change-about. I'm a delayed achiever.

Kathy's experience of consciously changing her way of perceiving life in general, with its consequent effect on her attitude toward mathematics, even to the extent that she became comfortable with estimation, is probably unusual. Yet one wonders if other people who are otherwise mathematically capable can be led to such change, at least in perceptions and attitudes towards mathematics, and whether they would then quite naturally become proficient estimators.

## Discussion and Conclusion

Combining data gathered through group-administered tests and scales and information gleaned from one-on-one conversations that allowed probing and exploration of individual answers proved to be a helpful way of characterizing affective factors that influence estimation ability. The first surprise generated by the interview data was the potential misclassification of students into good and poor estimators solely on the basis of group-administered tests. Two of the high scorers had qualitatively weaker skills and poorer understanding of numbers and operations than the other five. Similarly, one of the low-scoring students did quite well during the estimation part of the interview and attained a final ranking of sixth out of 17, higher than the two high scorers just described. In fact, 5 of the 10 students classified as poor estimators actually had fairly decent estimation skills but scored low because, in many cases, they simply needed to think about a problem a little longer than the timed test allowed before they were able to produce an acceptable estimation strategy. Thus, the five who remained in the lowest group were truly poor estimators. This reduced the final count in the two groups, but the students in the groups contrasted in the analysis were certainly better classified in terms of estimation skills, and there were more striking differences in responses on the affective measures than there would have been with the original two groups. Still, any conclusions drawn from such small numbers are admittedly tentative and conjectural in nature.

The resulting general profile of good estimators as people who have strong mathematics self-concepts, attribute success to ability, are unable to attribute failure to any factor because of inexperience with failure, and value mental computation and estimation skills was quite expected. The contrasting profile of poor estimators was also as expected. They had low mathematics self-concepts, attributed success to a great deal of time and effort on their parts, attributed failure to task difficulty and to not enough time and effort, and did not value mental computation and estimation very highly. The surprises came with the one student in each group who did not quite fit the profile. In Group 1, Stella's mathematics self-concept was clearly different for everyday mathematics and school mathematics, with her self-concept in school mathematics being quite low. She consequently attributed success in school mathematics to effort, rather than to ability. In Group 3, one of the poor estimators professed to enjoy mathematics, to have experienced success in past mathematics courses, and to possess a degree of confidence not felt by others in her group. The profiles, then, are only general classifications and do not necessarily characterize everyone in the group.

The final surprise came from the strikingly different paths that led some individuals to become good estimators. All five students could recount events and reactions to the events in their past that probably influenced their decisions, usually unconsciously made but, in one instance, quite consciously made, to develop good estimation skills. Without the impetus of courses or jobs that demanded estimation skills, formal instruction on estimation, strong parental encouragement, or a highly motivating factor, such as the need for estimation as a form of compensation, it seems doubtful that students will develop estimation skills on their own, even if they are capable of doing so. Instruction on estimation is now appearing in textbooks, however, and the need for developing estimation skills is recognized in many quarters (National Science Board Commission, 1982; Ralston, 1986), so perhaps it will no longer be necessary for individuals to develop these skills on their own.

The data analyzed in this study were obtained from either results of tests and scales or from interview protocols. The affective factors considered were from the realm of attitudes and beliefs, rather than emotions, yet there was one aspect of the study that provoked emotion on the part of many of the students and, in retrospect, should have been given more attention. I am referring to the students' reactions to the group-administered estimation test. The test items were necessarily timed in order to prevent actual computation. Several students later expressed discomfort with the test. They claimed they were often "just close to an answer," when the item was removed from the screen and replaced by another. Such an obvious interruption is certain to cause some degree of emotional response (Mandler, this volume, Chapter 1). The results of this response on student performance can be partially gauged by the necessary reclassification of some of the students after they completed the problems in an interview setting, without time limits. As instruction on estimation becomes more prominent in the mathematics curriculum, this point needs to be considered when planning appropriate assessment of estimation skills.

McLeod (this volume, Chapter 2) has argued that affective concerns need to be incorporated into curriculum development. Many of the affective characteristics of good estimators noted here are not typical of most school children. Instruction on computational estimation, as with topics like problem solving, cannot hope to be successful unless consideration is given to the affective factors influencing its development.

## *References*

Bestgen, B.J., Reys, R.E., Rybolt, J.F., & Wyatt, J.W. (1980). Effectiveness of systematic instruction on attitudes and computational estimation skills of preservice elementary teachers. *Journal for Research in Mathematics Education, 11*, 124–136.

Fennema, E., & Peterson, P. (1984). *Classroom processes, sex differences, and autonomous learning behaviors in mathematics: Final report to the National Science Foundation*. SE 8–8109077. Washington, DC: National Science Foundation.

Fennema, E., & Sherman, J. (1976). Fennema–Sherman mathematics attitude scales: Instruments designed to measure attitudes toward the learning of mathematics by females and males. *Journal for Research in Mathematics Education, 7*, 324–326.

Kloosterman, P. (1986). Attitudinal predictors of achievement in seventh-grade mathematics. In G. Lappan & R. Even (Eds.), *Proceedings of the Eighth Annual Meeting of the North American Chapter of the International Group for the Psychology of Mathematics Education* (pp. 244–249). East Lansing, MI: Michigan State University.

Lave, J. (in press). Problem solving as an everyday practice. In E.A. Silver and R.I. Charles (Eds.), *The teaching and evaluation of mathematical problem solving*. Reston, VA: National Council of Teachers of Mathematics, and Hillsdale, NJ: Lawrence Erlbaum Associates.

Maier, E. (1977, February). Folk math. *Instructor, 92*, 84–89.

National Science Board Commission: Report of the Conference Board of the Mathematical Sciences. (1982). *The mathematical science curriculum K–12: What is still fundamental and what is not*. Washington, DC: National Science Foundation.

Ralston, A. (1986). *The school mathematics curriculum: Raising national expectations* [Summary of conference]. Washington, DC: Mathematical Sciences Education Board.

Reyes, L.H. (1984). Affective variables and mathematics education. *Elementary School Journal, 84*, 558–581.

Reys, R.E., Rybolt, J.F., Bestgen, B.J., & Wyatt, J.W. (1980). *Identification and characterization of computational estimation processes used by in-school pupils and out-of-school adults* (Final Report, Grant No. NIE-79–0088). Washington, DC: National Institute of Education. (ERIC Document Reproduction Service No. ED 197 963.)

Rubenstein, R.N. (1985). Computational estimation and related mathematical skills. *Journal for Research in Mathematics Education, 16*, 106–119.

Threadgill-Sowder, J. (1984). Computational estimation procedures of school children. *Journal of Educational Research, 77*, 332–335.

Weiner, B. (1972). Attribution theory, achievement motivation, and the educational process. *Review of Educational Research, 42*, 203–215.

# 13
# Affective Issues in Teaching Problem Solving: A Teacher's Perspective

Verna M. Adams

It is encouraging that mathematics educators are talking about the relationship between mathematics learning and affect. Many teachers feel uncomfortable dealing with some of the psychological aspects of teaching mathematics. Even a teacher who is comfortable discussing a student's feelings of anxiety about mathematics in a one-on-one situation may feel inadequately prepared to deal with anxiety at the classroom level. Perhaps this has been an appropriate reaction to the situation. As many chapters in this book indicate, the relationship between affect and mathematics is complex, having many more components than just anxiety.

What does seem clear is that cognition and affect are interrelated (see the chapters in this book by Mandler [Chapter 1], Marshall [Chapter 4], and McDonald [Chapter 15]). This interrelationship implies that mathematics teachers should consider the effects of affective components of learning mathematics in planning effective instruction. In fact, attention to both affect and cognition may be needed in order to accomplish the goal of developing positive attitudes toward mathematics, a goal viewed by educators as an important outcome of teaching (Hart, this volume, Chapter 3). From interviews with elementary school teachers, Prawat (1985) suggested that "teachers who place equal emphasis on affective and cognitive goals" are more effective in "promoting positive attitudes toward others and toward the class as a whole" (p. 599) than teachers whose goal orientations are primarily cognitive or primarily affective.

As researchers investigate the relationship between affect and problem solving, how to plan effective instruction may become more clear to us. In the meantime, teachers are dealing with affective issues in one way or another. All teachers notice students' emotional reactions in the classroom, and sometimes classes seem to have an emotional identification. For the classroom teacher, the emotional tone in the classroom is as important as individual emotions. From this perspective, the cooler affective variables, such as beliefs and attitudes, seem as important as the hot affect that may occur during problem solving.

One environment for studying beliefs and attitudes is that of problem solving, because beliefs and attitudes play an important role in the emotion generated during the problem-solving process (Mandler, 1984; this volume, Chapter 1). Some of these beliefs about the learning of mathematics have been identified by Schoenfeld (1987) and are considered to be "important determinants of students' mathematical behavior" (p. 198). Student beliefs about mathematics and the learning of mathematics can have serious detrimental effects on problem solving.

In this chapter, I discuss some of my own observations about students' affective reactions to problem solving as I have moved toward using a problem-solving perspective in my own teaching. My comments on affect are based on discussions with students from Grade 7 to college and on self-reports from students that were part of written assignments that I called *problem-solving reports*. It seemed natural in these reports to have students reflect on their feelings and thoughts as they completed the assignment. In the first section of this chapter, I discuss these reports and some of the students' affective reactions to the problems used for the reports. I attempt to illustrate how the self-reports from students were used to modify instruction. The second section of the chapter deals with affective issues that the classroom teacher has to resolve in planning and implementing instruction.

## Problem-Solving Reports

Problem-solving reports were used as one type of assignment in a mathematics course for prospective elementary school teachers at San Diego State University. Students were given about 2 weeks to complete each assignment. They were allowed to work together or to get help, but they were required to do their own writing and were requested to report who gave them help. If the help was extensive, the student was supposed to ask the person about his or her thought processes while doing the work; the student was to record that information. After the reports were turned in to me, I read them and then discussed them with the class. The discussion usually included a summary of student affective responses as well as a report on the various methods students used to solve the problem. The purposes in discussing students' affective responses were to make students aware of the variability of the responses and to reassure them that their responses were not unusual. Examples of students' work were used to illustrate what I thought was good problem solving or good writing. Students were then given until the end of the semester to rethink and revise parts of the report. At that time, they were to turn in all of the writing completed for the course for a grade.

An example of a problem-solving assignment is given below. A major reason for making this assignment was to have students realize that they could ask questions about mathematics and investigate their own questions.

Problem:
Express the numbers 1 through 10 using 4 fours and any operation.
Assignment:
*Section I.* Solve the problem. Show your work.

*Section II.* List as many questions as you can think of that someone might ask about a problem of this type.

*Section III.* Investigate one of the questions on your list.

*Section IV.* Give a description of your thinking and feelings as you completed the above. What was your approach to the problem and to generating a list of questions?

This assignment was received favorably by the students, even though it created some frustration and anxiety. Some students asked the question, "Why use fours?" and looked for general forms for representing each number so that the fours could be replaced by any number.

In every assignment, students were asked to describe their thoughts and feelings as they completed the problem-solving report. For most students, this involved reflecting on the problem-solving process after it was completed. The process of explaining thoughts and feelings helps students to become aware of the role played in the problem-solving process by their emotions. One student reported that, for her, the value of writing about feelings was that it would help her when she worked with children: "By understanding how I feel I can be able to relate better to children as they approach math for the first time."

For the classroom teacher, the technique of having students write about feelings in relation to specific problems can be a rich source of information about differences between students' affective reactions to problems. For example, aesthetics seem to play a part in the solution process for some students. (See Silver & Metzger, this volume, Chapter 5, for a discussion of aesthetic judgments by expert problem solvers.) The following comments were made by a student trying to find the maximum number of lampposts that would be needed for a village with $n$ streets and a lamppost at each intersection:

The general rule I ended up with was

$$(n - 1) + (n - 2) + (n - 3) + \underline{\hspace{1cm}} + (n - n).$$

I wasn't satisfied with this answer: It was too long. I was looking for something nice and neat. I was pretty sure that this wasn't the final answer to the problem, but I couldn't figure out how to make a neater equation out of it.

The problem-solving reports gave information on each student, thus allowing for both an understanding of individuals and of the class as a whole. (See the work reported by Lester, Garofalo, & Kroll, this volume, Chapter 6, for a better understanding of the variability in students' responses to problem solving and the interrelatedness of cognitive and noncognitive factors.)

## Frustration

In the problem-solving reports, the most frequently occurring response was an expression of frustration. Frustration seems to vary with the problem, the student, and how the student has learned to cope with frustration. One student reported, "Each problem we get causes frustration, but each for a different reason." The open-endedness of the number-problem assignment in which students were to express the numbers 1 through 10 using 4 fours and any operation created insecurity and tension for several students. Here are three responses:

1. I prefer to do a problem where I can find one right answer, or one that has a limited range of answers. With this problem, I felt quite unsure of myself as I didn't know if there was 1,000 or 10 possibilities.
2. I think the "open-endedness" of this problem bothered me somewhat while I worked on it. I feel secure with "exactness" and definite solutions. I realized when answering my question for the third part that answering one question just seems to lead to another; you could go on forever.
3. My feelings while working this problem were initially apprehensive because I was suspicious over the fact of whether there really was a solution. When I found an operation for the first one, I gained confidence and went on to the next. While working on an operation for 2, I found operations for 8, 9, and 3. I began to enjoy this assignment as I tried different operations to get the solutions for the remaining numbers. By trial and error, I found the solutions and did not stop to rest my mind until the last numbers, 4 and 10, were solved. To get the more difficult ones, such as 4 and 10, I wrote multiples of four on scratch paper and studied them. This did not make it easy, but led me to try various combinations. My final feelings were of relief for being able to finish and of satisfaction for sitting down and facing the problem.

This last response is interesting for several reasons. One reason is that I get the impression that success at finding one solution enhanced the student's confidence and supported continued effort. After finding several solutions, the student's attitude shifted to enjoyment and sustained persistence. The final feeling of satisfaction for facing the problem is often expressed by students. One important source of positive feelings seems to be the satisfaction of having carried through on a task that was perceived as difficult.

Students sometimes begin to observe and report ways to deal with their frustrations. One student reported, "As I solved this problem for each number 1 through 10, I found that I was able to keep my frustration down by working on it a little at a time." Another student said, "I also tried a new strategy on this assignment by doing a small bit on it each day, so I didn't become frustrated. This was a much more pleasant experience than assignment one." Still another student reported using breaks to help change the student's mind set: "This problem required several breaks just so I could see each number in a fresh light."

Tension was not the only response to this number problem. Other students expressed liking the fact that they did not know if there was an answer. In fact, one student changed the problem to create more uncertainty. The student explored writing the numbers 1 through 10 using 4 threes and 4 fives.

I seemed to enjoy solving the problem using threes and fives more than I did using the assigned fours. I think this is because in the back of my mind I knew the fours would generate answers. However, I did not know this to be true using threes and fives. Therefore, I felt that I was discovering something. Somehow, I felt rewarded when the fives and threes all generated answers.

Another student reported that this particular problem was a challenge and expressed satisfaction at completing it. Something about the problem really intrigued the student.

Once I began working on this problem, I found it very difficult to stop working on this. I was having trouble finding the solution, but knew it was possible and I couldn't walk away from the problem until it was complete. I enjoy a challenge and I found this problem to be just that . . . . Although this was difficult for me, I enjoyed the feeling of accomplishment when I completed it. I found myself so involved with the assignment that I figured out more than one answer for each number.

## Modifying Instruction

The affective responses of students influenced decisions about the inclusion of topics in the discussion sessions for the problem-solving reports. One assignment, in particular, illustrates the influence of affective factors in the decision process. The assignment consisted of two problems that were written using nonsense words. The first problem was presented in paragraph form, and the second problem was presented in a form in which each sentence was on a separate line and there was a blank line between each sentence. The purpose in presenting the problems in this manner was to illustrate the influence of perception on how one might feel about solving a problem. Students were asked to choose one of the problems to solve and were asked to discuss why they chose that particular problem. Because the problems had been modeled after a problem in their textbook, the students were asked to find a similar problem and explain how the problems were alike. The first problem in the assignment was as follows:

On Whizzle Way in Waffle Wuz, there are 55 wugmuts. Twenty-one of the wugmuts are willets. Sixteen of the wugmuts are walmets. Twelve of the wugmuts are wimets. One half of the walmets are willets. One fourth of the walmets are wimets. One third of the willets are wimets. How many of the wugmuts are willets and walmets and wimets?

The second problem was the same as the first, except for the change in form and a change in the question: How many of the wugmuts are neither willets nor walmets nor wimets?

On the first day of class after distribution of the assignment, it was evident that many students found this assignment very frustrating. I interpreted the students' expressions of frustration as serious and modified my usual procedure of not discussing the solution to problems until after the assignment was completed. I asked students who were making some progress on the problem to explain what they were doing and discussed ways of dealing with the nonsense words.

When I read the reports, it was interesting to note that, for many students, letters, which were usually considered to be abstract by the students, were used to replace the key nonsense words. Only one student replaced them with words that made sense. How much this was influenced by the discussion on the second day of the assignment is not clear.

One student wrote in her report that, if she were the teacher, she would never give such an assignment to students. My interpretation of that comment was that she probably had some strong negative emotion while attempting to solve the problem and that the comment also reflected a belief about the teacher's role in the classroom. In reaction to her comments, I focused the discussion of the reports on what could be learned from such an assignment and what factors of the situation made the problem difficult.

For this assignment, I used judgments about students' affective responses to modify instruction both before students had completed work on the problem and after. In the first instance, I decided that the students needed more support than usual, and I provided support through class discussion of the problem. In the second instance, I attributed beliefs and difficulties to an individual as a result of what that individual wrote in the problem-solving report. I am concerned, however, about the accuracy of my interpretations of the students' responses and about the appropriateness of my instructional decisions. Will I be able to look to research in the future to help me improve both my interpretations and my decisions?

## Issues for the Classroom Teacher

Several issues about how to deal with affective reactions by students and how to plan instruction have emerged from my use of the problem-solving reports and other problem-solving activities. In this section, I raise the question of whether or not it is possible for instructional sequences to influence student affect. This is followed by a discussion of the need to align student expectations and teacher goals. Then I discuss how a teacher may be uncertain of the proper response to a student's particular emotional response to problem solving.

### Affective Objectives for Instruction

The general goal of generating positive affect seems to be too broad to deal with at the instructional level. When I first started having students complete problem-solving reports, the assignments were worded in very general terms, which did not generate much response from the students. As I created more specific objectives, the amount of information that students provided in the reports began to increase. The objectives that I developed were often related to self-regulation and to how students felt about the problem that was assigned. In working with sequences, for example, students usually were able to find the next term in the sequence, but it was more difficult for them to find the $n$th term for the sequence. This appeared to be a result of continuing to focus along the sequence rather than from a term in the sequence to its position in the sequence. To make students aware that they needed to shift their focus, I gave them the following assignment.

Problem:
   A. 1, 2, 3, 4, 5, 6, 7, ... , $n$
   B. 0, 1, 2, 3, 4, 5, 6, ... , ?
   C. 0, 2, 6, 12, 20, 30, 42, ... , ?
   D. 0, 1, 3, 6, 10, 15, 21, ... , ?
Assignment:
   *Section I.* Give a description of the relationships between the sequences above. Use words to describe what you would do to sequence A to get sequence B. Then describe what you would do to sequence B to get sequence C. Finally, describe what you would do to sequence C to get sequence D.
   *Section II.* Find the rules for the $n$th term for each sequence. Notice that the $n$th term for sequence A is given to you. It is just $n$. Use your observations from Section I to help you write the other rules.
   *Section III.* Give a description of your thinking and feelings as you worked.
   *Section IV.* Find a problem from the textbook or elsewhere that uses sequence D. Describe the way in which the problem uses the sequence.

The responses to Section III gave me feedback as to whether or not the students were making the shift in focus. Students who indicated frustration or who said that Sequence D was very difficult still had not made the shift in focus that I wanted. During class discussion, they expressed surprise at how easy it was to find the $n$th term if they made that shift. I then used the assignment to emphasize that the general rule was a statement of the relation between each term in the sequence and its position in the sequence (rule for the $n$th term). In contrast, finding the next number in a sequence could be found by looking for a relation between consecutive numbers in the sequence.

The cognitive and affective objectives for this instructional episode were closely related. The affective objective was to show students how the shift in focus could provide greater satisfaction through insight. A question that needs to be investigated is whether or not an instructional sequence such as this is effective in changing students' affective responses to a particular type of problem.

## Student Expectations and Teacher Goals

When I first started to try to incorporate more problem-solving into my teaching, I felt resistance by students to a problem-solving approach. In one class, there was a student who indicated that she had never before seen mathematics in the way that she was perceiving it in the class. We were focusing on how we think when we solve problems and on multiple ways of doing problems. She was working hard, asking good questions in class, and, in general, doing very well in the class. Sitting next to her was another student who complained bitterly about how I was teaching the class. He contended that he did not need any problem solving; he got plenty of that in other classes. He maintained that all he needed was for me to do lots of problems and if I would just give him enough examples, he would be able to do them.

Here were two people, sitting side by side, reacting in exactly opposite ways. My conclusion was that I was not the cause of student satisfaction or lack of satisfaction with the class. The first student was both surprised and pleased that the mathematics she was experiencing was unlike mathematics experienced in her past. The second student believed that mathematics should be something other than problem solving and that mathematics should be taught in a certain way.

For the classroom teacher, the issues are how to identify student expectations and how to resolve the differences between student expectations and teacher goals. If a teacher's description of the class is as explicit as possible, it seems to allow students to adjust their expectations somewhat. However, students' beliefs about mathematics and the teacher's role in the classroom seem particularly relevant to this issue. If students' beliefs are in conflict with the teacher's objectives, what can be done?

## Students Who Feel Happy When They Get a Wrong Answer

As one investigates students' feelings about problem solving, it is particularly troubling to be confronted with a situation in which a student who has not been very successful expresses great excitement at getting an answer, but the answer is wrong. The feeling of elation expressed by this student may have a good explanation. According to Mandler (1984), "the completion of interrupted actions and schemas is clearly a positive event" (p. 183). Getting an answer—any answer—when the student has been "stuck" and has not been confident that he or she could solve the problem at all may indeed be a moment of joy.

As a teacher, I do not know how to react. The problem has not been analyzed correctly. The conclusions are false. The student does not know enough to know the answer is wrong and has been open at showing feelings founded on incorrect work. What do I do? I do not want to embarrass the student. At the same time, I want to assist the student in acquiring better problem-solving skills. As a classroom teacher, I do not have the time necessary to work individually with the student in order to figure out how to help students who respond in this way. Would these students benefit from small-group instruction, or would small-group instruction just allow the student to avoid the issue of learning? (For an example of a student's reactions during small-group instruction, see the discussion of Marcus by Thompson & Thompson, this volume, Chapter 11.)

Another issue that seems to be difficult for the classroom teacher to resolve in the time that is available is mathematics anxiety. Students sometimes go to great lengths to avoid doing mathematics. On the surface, students may appear to be making considerable effort to pass the class. They will hire a tutor or see the instructor after class, usually to talk about their math anxiety. (For a discussion of off-task behaviors, see McDonald, this volume, Chapter 15.) It seems to take a long time for the student to realize the possibility that he or she is in control whether or not he or she learns mathematics. Sometimes students have beliefs about the nature of mathematics or about themselves as learners of mathematics that seem to interfere with their success in doing mathematics. Sometimes the student needs help with study skills, or the student has difficulty on a visual perceptual level. It seems to take considerable personal (one-to-one) contact to help students who express anxiety about doing mathematics.

## Summary

A major point of this chapter is that teachers are dealing with affective issues in the classroom and are making instructional decisions based on their interpretations of the affective responses of the student. As researchers study affect and problem solving, it is important not to exclude the teacher. Prawat (1985) reported that affective outcomes are frequently of primary importance in teachers' intentions and that elementary and secondary school teachers rely on student affective responses to evaluate the success of instruction. I have tried to illustrate, through examples from my own classroom, how student affective responses influence my instructional decisions.

Some of the issues with which teachers must deal seem to be closely related to student beliefs about mathematics, the role of the teacher, and themselves as learners of mathematics. The student's expectancies for the class need to be addressed by the teacher. If the student does not want or does not expect to get from the class what the teacher plans to provide, the student is usually unhappy.

Emotional reactions to problem solving may occasionally create situations that require the teacher to acquire new skills. The classroom teacher may also need to deal with the issue of how to provide students with support in overcoming mathematics anxiety. Mandler (1984) suggested that the behavior patterns of students with high anxiety differed from those of students with low anxiety. Reassurances given to a highly anxious student may counteract negative self-talk by the student. The same reassurances given to a student with low anxiety have the opposite effect. This suggests that the classroom may not be the appropriate place in which to deal with mathematics anxiety. Creating classrooms where anxiety is less likely to occur (see Cobb, Yackel, & Wood, this volume, Chapter 9, and Grouws & Cramer, this volume, Chapter 10) is certainly desirable.

I have tried to illustrate a method for collecting students' self-reports about affect as part of problem-solving reports. Having students report their reactions after completing a problem appears to be a technique that is manageable by the teacher; therefore, I hope that self-reports, in addition to other techniques, will be included in research studies so that we can improve our interpretation of the information collected after the problem solving has been completed.

# References

Mandler, G. (1984). *Mind and body: Psychology of emotion and stress*. New York: Norton.

Prawat, R.S. (1985, winter). Affective versus cognitive goal orientations in elementary teachers. *American Educational Research Journal, 22*, 587–604.

Schoenfeld, A.H. (1987). What's all the fuss about metacognition? In A.H. Schoenfeld (Ed.), *Cognitive science and mathematics education* (pp. 189–215). Hillsdale, NJ: Lawrence Erlbaum Associates.

# Part IV   Responses to the Theory

# 14
# The Study of Affect and Mathematics: A Proposed Generic Model for Research

ELIZABETH FENNEMA

During the last 15 years, much understanding has been gained about cognitions in mathematics. Starting with the content area of early number concepts, progressing through rational numbers and algebra, and culminating with problem solving by university mathematicians, scholars have been probing cognitions and determining how human beings think with mathematical ideas. This knowledge is beginning to be useful not only in understanding human thought processes, but also in the development of new paradigms for curriculum development (Carpenter, in press). This knowledge has grown and become useful because a relatively consistent research methodology has been used in relation to a fairly concise theoretical model of cognitive processing. Usually this cognitive science research methodology started with an explicit definition of some specific mathematics content; questions relating to this definition have been formulated; subjects were asked to think aloud (or self-report) as they answered the questions; their responses were studied for patterns of thinking; and, finally, the identified patterns were compared with the overall theory to see if the theory was supported or needed modification.

Research on affective variables, their development and relation to mathematical learning, has also used a self-report methodology. This methodology, although usually implemented with groups of learners, rather than with individuals, has resulted in certain kinds of knowledge that have also been useful in developing theories. However, the research on affective variables has not yet resulted in a cohesive picture, nor are there many implications that can be derived from this research area for teacher education or instruction. The purpose of this chapter is to explore the character of research dealing with affective variables and to discuss its strengths and weaknesses. I propose a theoretical model for work in the affective domain that might provide a cohesive framework for understanding the development of affective variables as well as their influence on mathematical learning. Such a model is generic in the sense that it can be useful in a variety of situations investigating diverse variables with many age groups. I also propose a blending of the traditional affective research methodology with a cognitive science methodology.

## Descriptive Work on Affect

During the years that our knowledge about cognitions in mathematics has been growing, other scholars have been attempting to understand how beliefs, emotions, feelings, and attitudes are related to the learning of mathematics. Although there are major problems with definitions (see Hart, this volume, Chapter 3; Leder, 1987), this broad spectrum of variables can be referred to as *affective variables* (see Krathwohl, Bloom, & Masia, 1964).

### *Traditional Methodology*

There have been many reported studies on affective variables. Some have been done in an attempt to understand the variables within the broad educational community (e.g., Shavelson, Hubner, & Stanton, 1976). Within the mathematics education community, the initial motivation for most of this work has been the desire to understand variables related to gender differences in mathematics. Using a traditional research methodology, variables thought to be important in explaining gender differences in mathematics were identified, and their components defined and operationalized into items combined into scales. Comparisons of scores on these scales have been made between groups of females and males, correlations computed between the variables and mathematics learning, and/or scores regressed onto achievement measures (see Meyer & Fennema, 1988, and Eccles, 1986, for reviews of this work).

This work has produced some interesting and useful knowledge. We know, for example, that confidence in one's ability to learn mathematics is correlated with mathematical achievement at about the .45 level (Fennema & Sherman, 1978); it is an important predictor of mathematics achievement, and it appears to be a better predictor of females' performance than of males' performance (Meyer, 1985). Perceived usefulness of mathematics is also correlated with achievement and predicts performance (Elliott, 1987). Subjects' beliefs about the usefulness of mathematics have been modified, and this modification has resulted in increased willingness to enroll in more mathematics courses (Fennema, Wolleat, Pedro, & Becker, 1981). Males report causal attributional patterns that are different than those reported by females (Wolleat, Pedro, Becker, & Fennema, 1980), and the pattern reported by males has been hypothesized to have a positive influence on learning.

Although some of the work on gender and affective variables has dealt with causal modeling (see Meyer, 1985; Eccles, 1986; Fennema & Peterson, 1985), most work has been descriptive, and focused on describing the differences between groups of people, usually males and females, on traits assumed to be relatively stable. The work has produced replicable results that have shown positive relationships between certain affective variables and achievement measures of various kinds. Indeed, this set of studies provided information that confirms the belief that attitudes toward mathematics are important. Moreover, the results of this research appear to be more significant than the studies produced over a number of years by previous investigators. (See Crosswhite, 1972, and Glennon & Callahan, 1968, for much more moderate conclusions about the relationship of affect and mathematics learning.) One reason for these more positive findings was the care with which the scales that measured the affective variables were developed.

Although other sets of scales were carefully developed at about the same time (Sandman, 1973; Dowling, 1978), many studies chose to measure affective variables with a common instrument—that is, the Fennema-Sherman (F-S) Mathematics Attitude Scales (Fenema & Sherman, 1976). The details of the development of these scales are important because the scales have been widely used and have resulted in many significant findings that enhance our knowledge about affective variables and mathematics. These scales had four important attributes: (a) variables to be measured were theory-based, (b) each attitude was carefully defined, (c) the scales were domain specific with respect to mathematics, and (d) they were developed analytically.

The F-S scales were developed as part of a National Science Foundation project aimed at identifying cognitive and affective variables related to gender differences in mathematics. They focused on the variables that had been hypothesized to be a cause of differences in achievement or in election of optional advanced mathematics courses. Nine affective variables were selected: (a) attitude toward success in mathematics, (b) mathematics as a male domain, (c) confidence in learning mathematics, (d) effectance (intrinsic) motivation in mathematics, (e) usefulness of mathematics, (f) mathematics anxiety, and (g) perception of mother's, (h) father's, and (i) teacher's attitudes toward one as a learner of mathematics. Each variable was embedded in a theoretical explanation of gender difference in mathematics (Fennema & Sherman, 1976). A Likert scale was developed to measure each of these variables.

Although the scales were theoretically independent, they were developed as a set. Many items were written for each scale. The items were validated by experts, and items for all scales were randomly distributed and administered to 367 high school students. Final selection of the items was made based on several criteria: (a) items that correlated highest with the total score for each sex, (b) items with higher standard deviations for each sex, (c) items that yielded results consistent with the theoretical construct of a scale, and (d) items that differentiated students who had elected to take advanced mathematics courses from those who had not elected such courses.

Each scale was specifically defined with respect to both the named variable and to mathematics. Consider, for example, the Confidence in Learning Mathematics Scale and its definition:

The Confidence in Learning Mathematics Scale (C) is intended to measure confidence in one's ability to learn and to perform well on mathematical tasks. The dimension ranges from distinct lack of confidence to definite confidence. The scale is not intended to measure anxiety and/or mental confusion, interest, enjoyment or zest in problem solving (Fennema & Sherman, 1976, p. 4).

Shavelson, Hubner, and Stanton (1976) had already proposed a theoretical model that showed the relation of confidence to learning, and there were data available that indicated that the theory was valid. Frieze, Fisher, Hansua, McHugh, and Valle (1978) and Dornbusch (1974) had indicated that females tended to underestimate their own intellectual activities (including mathematics) more than boys did. Thus, the theoretical underpinnings of this scale rested on the importance of confidence to everyone's learning and on the idea that there were gender differences in confidence. It was theoretically sound to hypothesize that confidence in mathematics might be an important variable that would explain, at least partially, gender differences in mathematics.

Each item in the scale was clearly related to the definition given, and validity was determined by experts who agreed that the item fitted the definition. One positive item was, "I am sure that I can learn mathematics." A negative item was, "I'm no good in math." The reliability of the scale as determined by the split half procedure was .93. Similar information is available for each scale. Although anyone can argue about the importance of what each scale measures, each definition is clearly stated, and the items that measure it were carefully selected.

These scales have been used in many reported studies. As a result, we now have a rich description of how males and females, from about age 9 to adulthood, differ with respect to the definitions of the nine scales. (A meta-analysis of results from these studies is currently underway.) Correlational studies using these scales have also indicated relationships between these variables and other affective variables, as well as between these variables and learning. In addition, we know the ability of these various scales to predict future election of mathematics courses and achievement in mathematics, and we know the stability of the measures over time (see Meyer, 1985; Pederson, Elmore, & Bleyer, 1985).

My purpose in detailing the development and use of these scales is not to praise any particular set of instruments but to describe how one research methodology, rigorously applied, has produced knowledge about affect and mathematics. This methodology is based in the paradigm of differential psychology, an approach to studying learner characteristics that has a long and successful history (see Leder, 1987, for a complete discussion). Knowledge obtained through these methods is almost exclusively descriptive, focuses on a trait or traits assumed to have some stability, and describes differences between groups and/or relationships between variables.

## Cognitive Science Methodology and Affect

Recently, descriptive work about affect has been studied using a modified cognitive science methodology, and this book includes many studies that have been done using this paradigm. Usually a learner is asked to think aloud about his or her feelings as mathematical problems are solved. Patterns of relationships between reported feelings and problem-solving behavior are sought. Some of the studies have been built on the carefully defined theory of Mandler (this volume, Chapter1). Consider, for example, the study reported by L. Sowder (this volume, Chapter 8), a beginning attempt to identify some of the perceptible affective variables or emotions that influence one's choice of strategy for solving routine word problems. He theorized that students should show some emotion when they reach a cognitive block or interruption in the process of solving a word problem. While students were solving problems in individual interviews, he watched carefully for signs of such emotion.

Lester, Garofalo, and Kroll (this volume, Chapter 6) used a similar methodology. During individual interviews in which students were solving problems using a think-aloud procedure, they watched for evidence of affective influence on problem solving. Silver and Metzger (this volume, Chapter 5) also used this methodology. As mature mathematicians solved problems, they watched for evidence that aesthetic monitoring would influence the processes selected for problem solving.

These studies focused on different affective variables. Sowder was looking for emotion; Lester and Garofalo were looking for beliefs and self-confidence; and Silver and Metzger were looking for aesthetics. The studies had many similarities, however. Each study was looking for the influence of affect on cognition. Each study attempted to describe individuals and to look for similarities and differences. Each study was based on Mandler's theory, which indicates that emotion or affect arises when a subject cannot see a clear path to problem solution and that this emotion should have an influence on mental processing.

Work such as that reported by Sowder, Lester and Garofalo, and Silver and Metzger represents a beginning stage in new descriptive work. From such work, we will begin to understand how affective variables influence mental processing; however, the need for careful definition of the variable under consideration is vital. It should be pointed out that work focused on cognitive processing of mathematical ideas starts with a complete definition of the mathematical idea being studied. Better definitions resulted in better and more useful information; thus, if we want to use a research methodology that probes influences of affective variables on cognitions, we must start with precise definitions of what affective variables we are looking for. This is not to say that we know what those precise definitions are. As we gain more information using a definition, that definition will be altered and ultimately improved. Without a beginning description of what we are looking for, however, little will be found.

Early research on affective variables resulted in nonsignificant results because it assumed that attitudes were global and undifferentiated. There was lack of definition, lack of clarity, and lack of connection to mathematics. It is possible to avoid making the same mistakes again as new ideas and research methodologies are employed. It is hoped that new researchers on affect will be clear about what is being studied, precise in definitions, and respectful of what has been learned previously.

## Affect and Educational Outcomes

Why should anyone study affect? Obviously, knowledge of human behavior and cognitions aid in our overall understanding of the human condition. Because emotions, feelings, and attitudes are part of human behavior and cognitions, to increase knowledge and build theories that describe human behavior and cognitions is, perhaps, sufficient reason to study affect. To mathematics educators, however, that reason should not be enough. What we need to know and understand is how affective variables relate to and influence important educational outcomes; therefore, a model is needed that could guide researchers to understand affective variables in relation to important educational outcomes. Before turning to such a model, however, it is important to clearly define educational outcomes, which are important when considering affect.

At least two important outcomes of mathematics education are related to affective variables: (a) beliefs, feelings, and emotions about mathematics, and (b) learning of mathematics. In other words, outcomes in both the affective and cognitive domains are important. This certainly is not a new idea. The two taxonomies of educational objectives proposed by Bloom (1956) and Krathwohl et al. (1964) were indications that objectives in both domains were a major focus of the curricula in most schools. Affective goals are still important to mathematics education. Two of five goals listed in the *Curriculum and Evaluation Standards for School Mathematics* (National Council of Teachers of Mathematics, 1988) are affective goals: Learning to Value Mathematics and Becoming Confident in One's Own Ability.

Many reports include a concern for better procedures with which to evaluate cognitive outcomes, which is also being addressed in the affective domain. The purpose of the Comprehensive Assessment and Instructional Planning project sponsored by The Ford Foundation and directed by Carol Tittle of City University of New York (personal communication, 1986) is to develop more effective assessment instruments in the affective domain. The long-term goal of the project is to use the results of both achievement tests and affective/motivational assessments to further instructional planning, increase student self-awareness of factors related to learning, and, as a result, forge a stronger link between teaching and learning in mathematics, particularly for women and minorities.

Another major purpose in studying affective variables is to determine their role in facilitating or hindering mathematical learning. For many decades, it has been believed that these variables are major influences on learning. Correlational studies have confirmed the relationship between affective variables and achievement, and other studies have shown that certain variables predict achievement (Fennema & Sherman, 1978; Meyer, 1985). In addition, teachers believe strongly that certain affective variables (usually those dealing with motivation or self-esteem) are a vitally important reason why children succeed or fail in mathematics (Fennema & Peterson, 1985).

Both outcomes of mathematical education (positive affect concerning mathematics and the learning of mathematics) must be taken into consideration as scholarly activity is planned. This suggests that the purpose of doing research with affective variables must be to increase understanding of the influence of these variables on mathematical learning as well as the development of these variables in learners. In order to plan research that will achieve these purposes, a theoretical model that hypothesizes the relationship between affective variables and these important educational outcomes is essential. Such a model gives direction to research and provides the common ground where many studies can be linked. Only when many studies share a common focus can enough information be gathered that will truly help in understanding the role of affective variables in mathematics learning. A previously proposed model can serve as a guide in the development of a new model.

## The Autonomous Learning Behavior Model

The Autonomous Learning Behavior (ALB) Model was first proposed by Fennema and Peterson (1985) as a possible explanation of gender differences in mathematics. Such differences are some of the most pervasive and persistent educational inequities that exist (Meyer & Fennema, 1988). The purpose of the ALB model was to place consistently found gender differences into a perspective that would yield relevant research hypotheses. This model provides a general paradigm from which a more generic model can be derived.

In order to explicate the ALB model, I present a brief description of gender differences in mathematics. Following this, I present the characteristics of each component in the model and a theoretical rationale for the model and its components.

### Gender Differences in Mathematics

Although there has been much research about gender differences in mathematics and many interventions have been developed to alleviate the differences over the last 15 years, there is still powerful evidence that males achieve at higher levels in mathematics than do females, particularly in tasks of high cognitive complexity, such as true problem solving. Although it has been believed that such differences do not appear until early adolescence, in a recent first-grade study that used a specially developed test of problem solving, boys performed significantly higher than did girls (Fennema, Peterson, & Carpenter, 1987). Although there is some disagreement as to the importance of the differences in achievement, it is safe to conclude that, when significant differences are found, males score higher. It appears that as a test's ability to measure real problem solving increases, so does male performance over female performance. There are also gender differences in three important affective variables. Males indicate more confidence in their ability to learn mathematics, report higher perceived usefulness, and attribute success and failure in mathematics in a way that has been hypothesized to have a more positive influence on achievement (see Meyer & Fennema, 1988, for a more complete discussion).

In addition to differences between the genders in the three affective variables just cited and in mathematics learning, there are also differences in how teachers interact with and treat girls and boys. Boys interact more with teachers than do girls. Girls have many more days in which they do not interact at all with the teacher than do boys. Teachers initiate more contacts with boys than with girls. Boys receive more discipline contacts as well as more praise. Teachers respond more frequently to requests for help from boys than from girls and tend to criticize boys more than girls for the academic quality of their work (Fennema & Peterson, 1987). Grieb and Easley (1984) report that teachers allow some boys to exert their independence overtly by refusing to learn specific algorithms.

In summary, there are gender differences in mathematical problem solving, in affective variables important in learning mathematics, and in how teachers interact with and treat learners. In each case, males show more positive outcomes and influences than do females.

## The Model

The ALB model is presented in Fig. 14.1. One should keep in mind that this model was designed to help in understanding gender differences in mathematics learning; thus, the somewhat isolated factors described in the preceding section became the major elements of the ALB model. As this was done, it became clear that a mediating factor between what was known about affective variables, classroom treatment, and educational outcome was missing; thus, the component *autonomous learning behavior* was inserted. The validity of this factor could then be studied in a variety of ways.

At the right side, or outcome end, of the ALB model is performance on high-level cognitive skills such as problem solving which involves finding the solution to a problem that one does not already know how to solve. One has to go beyond the mental processes one already possesses and do something that is different than what was done before. In order to solve a problem, then, it is necessary to think independently or autonomously. McCombs (1986) concurs with such a definition of ALB when she says that independent thinking requires "self-directed, self-regulated and autonomous information processing activities" (p. 2).

Fennema and Peterson (1985) hypothesized that, in order to develop the skills required to be an independent problem solver, one must participate in activities that develop autonomy (i.e, autonomous learning behaviors [ALBs]). The ALBs that develop autonomy include thinking independently about problem solving, choosing to do it, persisting at it, and achieving success in it. In other words, one learns to solve problems by solving problems. The more one engages in problem-solving activities, the better one learns problem solving. These activities can be done either in or out of school and lead one to be an independent or autonomous thinker. It should be pointed out that it is possible to participate in ALBs in groups as well as by oneself. The critical component is that the activity forces (or encourages) one to engage in mental activity that goes beyond what one has done before and to do this independently.

Although the ALB element in the model was originally only a hypothesis, some research indicates that participating in these behaviors is important in learning. Gustin (1982) reported that 23 exceptionally talented mathematicians

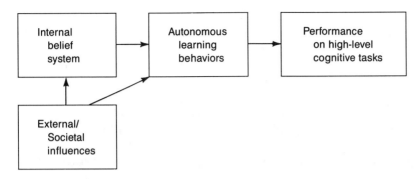

FIGURE 14.1. The autonomous learning behavior model.

described themselves as independent learners. According to Gustin, the opportunity to work independently during early school years was extremely important to these mathematicians. Having teachers who did not supply answers appeared to facilitate ALB development. Grieb and Easley (1984) argue strongly that independence in learning is essential to achieving a conceptual framework in mathematics that enables one to continue to use mathematics or to do high-level cognitive tasks. They contend that memorization and independent thinking are typically exclusive of one another. An inspection of biographies of outstanding mathematicians shows clearly that they have traits of independence and persistence and that they choose to do those mathematical activities that are mentally stimulating (Helson, 1980).

In summary, in order to solve mathematical problems, one must be able to think independently. One must choose to work on challenging tasks, figure out solutions independently, and persist until the task is satisfactorily completed. Learners learn what they do. Those learners who participate in more ALBs will learn to be more autonomous and, thus, will be better problem solvers than those learners who participate in fewer ALBs.

As indicated by the model, an internal belief system influences the participation in ALBs. Although there are probably many elements in this internal belief system, data indicate that, when considering gender differences, three interrelated, affective variables are of importance: (a) confidence in one's ability to do mathematics, (b) perceived usefulness of mathematics, and (c) causal attribution in mathematics. An internal belief system influences whether or not one is willing to become mentally involved in those ALBs that enable one to learn to be autonomous. (See Fennema & Peterson, 1985, for a more thorough discussion of these variables, and McCombs, 1986, for an information-processing discussion of the relationship of affective variables and development of autonomy.)

*External and societal factors* influence participation in ALBs both directly and indirectly, through the development of the internal belief system. Although there are many important external factors, I discuss only those that have the most direct influence on the learning of mathematics and those that can be changed by educational interventions (i.e., classrooms). Also, because most opportunities to participate in mathematics occur in classrooms, it is in classrooms where learners have the most opportunity to participate in ALBs. One very important factor in the classroom is the teacher. The things the teacher says and does, the beliefs and expectations held by the teacher, and the activities in which learners are expected and encouraged to participate are all ways in which teachers influence students' internal beliefs and actual participation in ALBs. The way a teacher treats a learner is also an important influence on the learner's internal belief system.

In summary, the ALB model demonstrates the relationship between affective variables, classroom influences, and important outcomes of mathematics education. Participation in ALBs acts as a mediator between affect, teachers, and outcomes. In particular, the ALB model is useful in understanding the relationship between affect and gender differences in mathematics. Several studies support the validity of the ALB model (Fennema & Peterson, 1986; Fennema & Peterson, 1987; and Peterson & Fennema, 1985).

Others working within the area of gender differences in mathematics have pursued the relationship between affect and outcome in different ways. Eccles (1983) and Meyer (1985) utilized path-analysis techniques to show how affective variables influenced educational outcomes. Eccles explored the relationship between affect and the choice of mathematics courses in secondary school, whereas Meyer used LISREL to explore the relationship between causal attributional patterns, confidence, and mathematics achievement. We do know a great deal about the influence of affect on gender differences in mathematics.

## A Generic Model

What does the research on gender-related differences in mathematics and affective variables suggest about designing a generic model for research in the affective domain? Why has that work been successful, at least to some degree, in exposing the influence of affect on educational outcomes of females? First, the selection of relevant affective variables was based on theory. The fact that the variable *confidence in learning mathematics* has provided significant results is not a matter of luck. A large body of psychological knowledge indicated that the general variable of confidence or self-esteem was important to females; thus, there was strong suggestion that using such a variable would produce significant results when used with mathematics learning. Second, in contrast to early research on attitudes, the variable was clearly and precisely defined. Third, the variable was translated and clearly defined with respect to mathematics. Although there might be some general propensity for a particular affect to be manifested throughout a person's behavior, it appears that affect may operate very differently in various content domains. Just because one is very confident of one's ability to understand poetry, for example, does not mean that one is just as confident with respect to mathematics. Fourth, the variable of concern was measured using a method that attested to its validity and reliability. Fifth, the outcomes of concern were clearly specified in relation to mathematics; the outcomes were achievement, participation in high-school mathematics courses, and/or beliefs about oneself and mathematics. Each outcome was more precisely defined by specifying levels at which it might be exhibited. When the outcome of concern was achievement, for example, it was tied to different cognitive levels of processing in mathematics. Sixth, it was hypothesized that the clearly defined variables were connected to the defined outcomes through certain mediating variables. In the ALB model, the mediating variables included the ALBs of choice and persistence. The affective variables of confidence, attributional style, and perceived usefulness of mathematics influenced the willingness of a learner to choose to do mathematics and to persist at mathematical tasks, which influenced the learning of mathematics.

Kloosterman (1984) has proposed a model in which affect (attributional style) influences learning through the mediating variable of participation in cognitive activities. In another model, McCombs (1986) suggested that one's beliefs about oneself, such as self-esteem, influence how that person is willing to interact with any information; thus, those beliefs influence learning.

How does the descriptive work based on a cognitive-science methodology fit into this discussion? In much of the work reported in this volume, the investigators have attempted to identify some affect that is expressed overtly as subjects solve mathematical problems. This work also appears to fit the generic model. Consider the theory. According to Mandler, when a person reaches a point during problem solving when he or she does not know what to do, a blockage has occurred and affect is the result. This affect then influences how one proceeds toward a solution to the problem. Perhaps this is an extension of the ALB model. Consider the following scenario:

Anne (a beginning first-grade student) is asked to solve this problem: "Jane has 4 pennies and Bill has 9 pennies. How many more pennies does Bill have than Jane?" To an adult, this is a simple subtraction problem. However, the problem is very difficult for Anne. Previously, she has solved only simple subtraction problems in which one set is separated from another set. This problem, which involves comparing one set with another, presents a challenge to her and is a true mathematical problem. She is blocked because her previous solution strategies do not work. However, Anne persists at trying to find a solution. She tries using her fingers but that does not help. She sits and thinks for awhile and decides to model each set with different colored counters. After more thinking (when not too much overt behavior is apparent), she quickly lines up one set, lines up the second set beneath the first set, counts how many stick out beyond the first set, and says "5."

Certainly, Anne exhibited an ALB. She persisted until she figured out a way to model the problem, and then the solution was easy to find. According to Mandler, when the child was blocked and did not know what to do next, whatever affect was felt (and it is unclear what affect was felt in the scenario and often unclear to an observer in real life) influenced Anne as she continued in her problem solving, and she eventually had a successful problem-solving experience. The ALB model does not specify as precisely as does Mandler what takes place mentally and emotionally, but the ALB model does suggest that affect influences persistence, which, in turn, does influence the learning of mathematics. It suggests that internal belief or affect was at least one reason why Anne was willing to persist in her problem-solving behavior.

The new work on affect, which uses a cognitive-science perspective, is an attempt to explore the influence of various affective factors on mental activities or cognitions that take place. In much the same way that early work on information processing and its translation to the mental processing of mathematics was conceived and carried out, interest is now focused on understanding the influence of *mental processing on affect*. It has taken at least two decades to gain enough knowledge about learners' mathematical cognitions to understand how that knowledge should be applied to decisions about mathematical curricula (see Fennema, Carpenter, & Peterson, 1986). It is hoped that it will not take that long before we can understand affect and mental processing enough to develop implications for curricula.

## A Generic Model for Research on Affect and Outcome

A generic model for relating affect and outcome might look like the one shown in Fig. 14.2. Each component encompasses a broad range of phenomena and can be refined in relationship to the researchers' concerns. At certain times, one element or the other might provide a total focus for a program of study. For example, to really understand affect, we may need to devote 10–15 years of research solely within a cognitive-science perspective. Such an approach has proved profitable for those working on children's cognitions. However, we should always keep in mind that our long-term goal as mathematics educators is to use the model to improve the teaching and learning of mathematics.

It appears that we have a lot to learn from previous work with affective variables. There are at least two areas in which traditional methodology can provide guidance: the first is in making decisions about which variables one should study, and the second is in the care with which the work is done. It is clear that thoughtful planning must precede scholarly work if it is going to result in replicable knowledge. We must define what variables we are talking about in relation to mathematics, and we must measure them carefully. We should have hypotheses about their relation to mathematics and learning. It is significant that the work within the area on children's cognitions has resulted in important knowledge because of careful modeling or hypothesizing. Work within the affective domain should follow suit.

In addition, perhaps the two theoretical perspectives (differential and cognitive approaches) could complement each other. For example, we might decide that pervasive, affective traits that have been studied before (such as confidence in learning mathematics) might be helpful in studying an individual emotion during problem solving. Selecting high-and low-confidence students might maximize the possibility of overt manifestation of emotion.

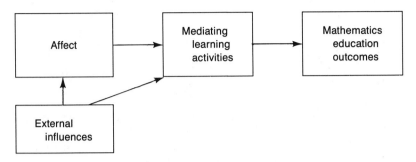

FIGURE 14.2. A generic model for relating affect and outcome.

Certainly, scholarly work that results in understanding of affect and mathematics is essential, but it is also very difficult work to perform. Although overt manifestations of affect can be described, the important affect lies deeply buried within each one of us. In our society, we are taught to conceal many of our emotions, beliefs, and feelings, even from ourselves. How to identify these feelings without invading personal privacy is not clear. Do we even have the right to try? Because this internal affect does influence mathematics learning, it seems that we should keep struggling with research on affect until most students learn mathematics and feel positive about themselves as learners of mathematics.

## References

Bloom, B.S. (Ed.). (1956). *Taxonomy of educational objectives: Handbook 1. Cognitive domain*. New York: David McKay.

Carpenter, T.P. (in press). Teaching as problem solving. In R.I. Charles & E.A. Silver, *The teaching and assessing of mathematical problem solving*. Reston, VA: National Council of Teachers of Mathematics and Hillsdale, NJ: Lawrence Erlbaum Associates.

Crosswhite, F.J. (1972). *Correlates of attitudes toward mathematics* (National Longitudinal Study of Mathematical Abilities, Report No. 20). Palo Alto, CA: Stanford University Press.

Dornbusch, S.M. (1974). To try or not to try. *The Stanford Magazine, 2*, 50–54.

Dowling, D.M. (1978). *The development of a mathematics confidence scale and its application in the study of confidence in women college students*. Unpublished doctoral dissertation, Columbus: Ohio State University.

Eccles, J.S. (1983). Expectancies, values, and academic choice: Origins and change. In J. Spence (Ed.), *Achievement and achievement motivation* (pp. 75–146). San Francisco: W.H. Freeman.

Eccles, J.S. (1986, March). *Gender roles and vocational decisions*. The presidential address delivered to the American Psychological Association, Washington, DC.

Elliott, J.C. (1987). *Causal attribution, confidence, perceived usefulness, and mathematics achievement of nontraditional female and male college students*. Unpublished doctoral dissertation, University of Wisconsin, Madison.

Fennema, E., Carpenter, T.P., & Peterson, P.L. (1986). *Teachers' decision making and cognitively guided instruction: A new paradigm for curriculum development*. Paper presented at the Seventh Annual Psychology of Mathematics Education Conference, London, England.

Fennema, E., & Peterson, P.L. (1985). Autonomous learning behavior: A possible explanation of gender-related differences in mathematics. In L.C. Wilkinson & C.B. Marrett (Eds.), *Gender-related differences in classroom interactions* (pp. 17–35). New York: Academic Press.

Fennema, E., & Peterson, P.L. (1986). Teacher-student interactions and sex-related differences in learning mathematics. *Teaching and Teacher Education, 2*(1), 19–42.

Fennema E., & Peterson, P.L. (1987). Effective teaching for girls and boys: The same or different? In D.C. Berliner & B.V. Rosenshine (Eds.), *Talks to teachers* (pp. 111–125). New York: Random House.

Fennema, E., Peterson, P.L., & Carpenter, T.P. (1987). *Teachers' beliefs about girls, boys and mathematics.* Manuscript in progress.

Fennema, E., & Sherman, J. (1976). Fennema-Sherman mathematics attitude scales: Instruments designed to measure attitudes toward the learning of mathematics by females and males. *Journal for Research in Mathematics Education, 7*(5), 324–326.

Fennema, E., & Sherman, J. (1978). Sex-related differences in mathematics achievement and related factors: A further study. *Journal for Research in Mathematics Education, 9*(3), 189–203.

Fennema, E., Wolleat, P., Pedro, J., & Becker, A. (1981). Increasing women's participation in mathematics: An intervention study. *Journal for Research in Mathematics Education, 12*(1), 3–14.

Frieze, I.H., Fisher, J., Hansua, B., McHugh, M.C., & Valle, V.A. (1978). Attributions of the causes of success and failure as internal and external barriers to achievement in women. In J. Sherman & F. Denmark (Eds.), *Psychology of women: Future directions of research* (pp. 519–552). New York: Psychological Dimensions.

Glennon, V.J., & Callahan, L.G. (1968). *A guide to current research: Elementary school mathematics.* Washington, DC: Association for Supervision and Curriculum Development.

Grieb, H., & Easley, J. (1984). A primary school impediment to mathematical equity: Case studies in role-dependent socialization. In M. Steinkamp & M.L. Maehr (Eds.), *Women in science: Vol. 2. Advances in motivation and achievement* (pp. 317–362). Greenwich, CT: JAI Press.

Gustin, W. (1982). *Learning to become a mathematician: The development of independence.* Paper presented at the annual meeting of the American Educational Research Association, New York.

Helson, R. (1980). The creative woman mathematician. In L.H. Fox, L. Brody, & D. Tobin (Eds.), *Women and the mathematical mystique* (pp. 23–54). Baltimore, MD: Johns Hopkins University Press.

Kloosterman, P.W. (1984). *Attribution theory, learned helplessness, and achievement in ninth-grade mathematics.* Unpublished doctoral dissertation, University of Wisconsin, Madison.

Krathwohl, D.R., Bloom, B.S., & Masia, B.B. (1964). *Taxonomy of educational objectives: Handbook 2. Affective domain.* New York: David McKay.

Leder, G.C. (1987). Attitudes toward mathematics. In T.A. Romberg & D.M. Stewart (Eds.), *The monitoring of school mathematics* (pp. 261–277). Madison, WI: Wisconsin Center for Education Research.

McCombs, B.L. (1986, April). *The role of the self esteem in self-regulated learning*. Paper delivered at the annual meeting of the American Educational Research Association, San Francisco.

Meyer, M.R. (1985). *The prediction of mathematics achievement and participation for females and males: A longitudinal study of affective variables*. Unpublished doctoral dissertation, University of Wisconsin, Madison.

Meyer, M.R., & Fennema, E. (1988). Girls, boys, and mathematics. In T.R. Post (Ed.), *Teaching mathematics in grades K-8: Research-based methods* (pp. 406–425). Newton, MA: Allyn & Bacon.

National Council of Teachers of Mathematics. (1988). *Curriculum and evaluation standards for school mathematics*. Reston, VA: National Council of Teachers of Mathematics.

Pederson, K., Elmore, P., & Bleyer, D. (1985). *Affective and cognitive mathematics-related variables: A longitudinal study*. Carbondale, IL: Southern Illinois University.

Peterson, P.L., & Fennema, E. (1985). Effective teaching, student engagement in classroom activities, and sex-related differences in learning mathematics. *American Educational Research Journal*, *22*(3), 309–335.

Sandman, R. (1973). The development, validation, and application of a multidimensional mathematics attitude instrument (Doctoral dissertation, University of Minnesota). *Dissertation Abstracts International*, *34*.

Shavelson, R.J., Hubner, J.J., & Stanton, G.C. (1976). Self-concept: Validation of construct interpretations. *Review of Educational Research*, *46*(3), 407–441.

Wolleat, P.L., Pedro, J.D., Becker, A.D., & Fennema, E. (1980). Sex differences in high school students' causal attributions of performance in mathematics. *Journal for Research in Mathematics Education*, *11*(5), 356–366.

# 15
# Psychological Conceptions of Mathematics and Emotion[1]

BARBARA A. McDONALD[1]

Mathematics and emotion. Isn't mathematics supposed to be nonemotional? Isn't mathematics the essence of pure, objective, precise, untarnished thought? As educators, however, we know that many students have emotional reactions, often negative ones, to mathematics. Furthermore, it is my contention that every thought has an emotional component. The connection between cognitive and emotional processing (if they are separate processes at all) is fundamental. The participants in the Conference on Affective Issues in Mathematical Problem Solving were concerned with mathematics and emotion because they have recognized that learning is not all cognitive, that cognition and emotion are intertwined. We are attempting to understand the connection. The conference gave us an opportunity to discuss the cognitive-emotional interaction, define it for mathematics, understand the process, suggest methods for studying it, and find examples that support our views.

Discussions at the conference were guided by the theoretical perspective of George Mandler (1975, 1984, this volume, Chapter 1). Our task was to use his theory to explain mathematics learning and emotional reactions. In doing so, many issues arose. Two issues that emerged and that were discussed throughout the conference have stayed with me and bear repeating as a framework for my discussion in this chapter. The first of these issues concerns the relation between emotions and information processing: What is the nature of the process by which emotions interact with information processing in the learning of mathematics? At the conference, this question generated discussions about individuals as they confront difficult learning experiences, discussions about which emotions are involved, and discussions about how emotions change the learning and memory for mathematics. The second issue concerns the belief systems and attitudes that predispose students to differentially respond to mathematics learning. Attitudes and beliefs were presented as cool forms of emotional reaction, and it was questioned whether or not this direction of inquiry was useful to pursue. At the end of the conference, we were left with these additional issues: What types of emotional reaction are we referring to and how shall we study them? Furthermore, of what practical value will our investigations be?

It is my intention in this chapter to respond to the issues presented at the conference by discussing my understanding of Mandler's theory and by adding my own elaborations of his ideas. In addressing the two issues that seemed most central to the conference, I put the question of cognitive and emotional processing interaction into a developmental as well as a social framework. In my analyses of both of these frameworks, the concept of self-esteem is critical. I suggest that only by considering both cognitive and emotional processing can we understand learning.

## Mandler's Theory

There are many models of emotion and many accounts of the cognitive-emotional interaction (see Bower, 1981; Lazarus, 1984; Tomkins, 1979; Zajonc, 1980). However, none are better suited to helping us understand the cognitive-emotional interaction in learning than the theory of Mandler. The reasons for his theory's applicability to learning are many. Mandler is an information-processing theorist. His theory concerns process, thus it allows us to see what is happening during learning. His cognitive model is more than a network of nodes; he takes into account the whole person. To Mandler, emotion is hot and takes place within the physical realm of the body. He emphasizes the expectations of the individual in conjunction with task interruption, a common occurrence in learning. The consequences of this task interruption can induce the individual to process information that is unrelated to the topic he or she was pursuing before the interruption occurred. According to Mandler's information-processing theory, at any moment there is a great deal of activity occurring, much of it preconscious or unconscious. All of these theoretical components lead to a complex representation of the individual and his or her reactions to events. Mandler's theory thus provides a comprehensive and well-integrated view of the person.

Mandler considers emotional experience to be the result of a combination of cognitive analyses and physiological responses. In his view, interruption of purposeful, planned, cognitive task processing gives rise to elevated autonomic nervous system (ANS) activity, which is an undifferentiated physiological reaction unless it has meaning to the individual, in which case it is processed cognitively, like any of the other contents of experience. When this ANS activity is analyzed cognitively, it is reflected in the emotional response of the individual. Interruption causes this physiological and cognitive reaction because it is contrary to expectation. According to Mandler, human beings do not like contrary experiences; they prefer predictable, controlled experiences. The responses Mandler is talking about happen so fast that we are not necessarily aware of this process.

The cognitive analyses accompanying elevated ANS activity do not need to be conscious and probably rarely are conscious. By including in his theory both conscious and unconscious processing of information, Mandler is able to capture the important aspects of the cognitive-emotional interaction, because it would be difficult to discuss emotions without referring to the unconscious. Furthermore, it permits us to consider the question, "What part of the process is conscious and what part is unconscious?" rather than the question, "What is the cognitive component and what is the emotional component?"

The cognitive analysis of the physiological reaction, although not necessarily conscious, does involve extensive processing that includes the attitudes and beliefs of the individual. According to Mandler, the cognitive analyses that accompany ANS arousal depend on the interpretation of the event by the person, and this interpretation is dependent on the values that are held. It is here that Mandler acknowledges beliefs and attitudes as having an impact on emotional expression or as causing the interruption in the first place. I shall consider this notion later in this chapter.

With an understanding of the individual, as provided by Mandler, we can see that mathematics learning and emotion need not be restricted to scenarios that are simple, or cut and dried, such as (a) task processing, (b) mistake, (c) emotional reaction, and (d) returning to the task or not. Mandler's analysis allows us to understand what is happening in much more complex and true-to-life scenarios, such as when a person becomes involved in a task, makes an error, and, instead of going on and trying again, wanders off into fantasies of his or her own incompetence (which may consist of or lead to self-pity or may be vigorously defended by ego-defense mechanisms). To me this is the far more interesting type of scenario to understand in learning. In my own work, this perspective allows speculation about how ideas are selected by the cognitive system for further processing. It makes it possible to understand that the choice of topic to be processed and the direction in which processing will proceed are dependent on the connection between cognitive and emotional processing and that this often happens unconsciously. Thus, the richness of the interaction between cognitive and emotional processing is easier to understand using the analysis provided by Mandler.

I share Mandler's interest in task interruption as a source of emotion. I am particularly interested in how task processing, once interrupted, can become so non-task oriented, why self-evaluations about performance are so frequently invoked when people get off their task, and how the thoughts that accompany these off-task occurrences are so frequently defensive. In the following section, I begin with an exploration of the reasons why people do not necessarily return to the task following interruption. This issue provides a starting point in understanding how information is processed cognitively and emotionally.

## Time Spent Off Task

The distinction between on-task and off-task behavior is just as complex as that between cognitive and emotional processes.[2] It is often counterproductive to get off task, yet it is very common. Why don't many students return to the task when they have been interrupted by error or for some other reason? The tendency to get off task has been discussed in terms of a fixed-capacity cognitive resources tradeoff model by some researchers who study information processing. In the work of Kanfer and Ackerman (1988) and Revelle (1987), attention is conceptualized in terms of resources available to put toward the task. These resources are dependent on the nature of the task, the difficulty of the task, the amount of effort the individual chooses to allocate to the task, and the attentional demands associated with the amount of practice on the task. These models conceptualize the importance of cognitive, motivational, and emotional resources in task performance.

Kanfer and Ackerman (1988) have studied learning behavior in a simulated air-traffic-control task. In this study, they investigated the effects of goal-setting instructions on learning performance. Presumably, goal setting requires attentional effort. Their experimental procedures allowed them to monitor time spent on task, and they found that attention to the task was not consistent for all students. In particular, they found that the imposition of a goal early in practice interfered with learning for high- and low-ability students. Specifically, persons low in ability engaged in more self-evaluative activity when given an explicit goal. After students had an opportunity to understand the task, however, the performance goals lead to greater learning among low-ability students than among high-ability students. These results suggest that self-evaluation may be positive or negative, depending on students' understanding of the task. Although low-ability students continued to engage in more self-evaluation, these activities appear to have enhanced effort and learning once the task was understood.

The work of Revelle and his colleagues (Anderson & Revelle, 1982; Humphreys & Revelle, 1984) also addresses the issue of attention diverted from the task during task performance. They have studied information processing using a very specific division of cognitive tasks—those tasks that require sustained information transfer, short-term memory, and long-term memory. They also use the motivational characteristics of effort and arousal, and use the personality characteristics of impulsivity and anxiety. The personality variables are used to explain certain tradeoffs in the acquisition of cognitive skills; anxiety has been responsible for time spent off task. They found that high-anxiety subjects allocated fewer resources to the experimenter-defined task than did low-anxiety subjects. According to Leon and Revelle (1985) and Sarason (1975), anxious subjects spent time in evaluating self-esteem, which is perhaps less threatening than engaging in the experimenter-defined task.

Kanfer and Ackerman also found indications of anxiety in their low-ability subjects, but some of their other data make the situation more complex. Although these students indicated their anxiety by frequently checking their scores, they perceived the task to be less demanding than did the high-ability students. Presumably, the task is confusing for low-ability students because they are not doing well; yet they do not indicate that they are confused, at least in the direct manner of reporting it. Kanfer and Ackerman suggest that these students go off task quickly because of anxiety, and that this delays their understanding of the task (R. Kanfer, personal communication, March 14, 1988). An additional explanation could be that their reporting is defensive. It would appear from this type of data that a more in-depth analysis of personality variables is in order.

It is not surprising that students spend time in off-task behaviors, because that is where emotion dominates the picture. The student may not consider the time to be wasted. Perhaps checking scores gives the student an action to perform, something to do. Maybe it is understandable that a person facing a threat to his or her ego will check to confirm performance results. In these cases, individuals are directing attention away from the task and toward information about their performances, which, if the student is failing, must have a negative impact on self-concept and certainly has a negative effect on learning. Clifford (1984) conceptualizes response to failure to be a combination of behavioral and affective actions or feelings. In her analyses, affective responses to failure, such as embarrassment, can be made in addition to or in spite of such behavior as giving up and not persisting. If the emotional reactions to failure or task interruptions are extreme, it is understandable that students go off task; attention is not directed toward the task. Because of the emotional nature of the interruption, the self-concept is implicated, and a search for an explanation of performance (attribution) and a possible defense for the experience is attempted. These experiences are emotional reactions to interruptions in the task. To return to Mandler's theory, interruptions are cause for emotion, and when the cause of interruption is an error or some failure or misunderstanding, it is not a trivial experience for the student.

Mathematics can be a school subject that produces many opportunities for task interruption. The emotional responses occur because students do not have the same feelings of control that they might have in other subjects. Mathematics is different from most verbally oriented subjects. It relies on a different symbol system than other school subjects. Furthermore, from the way mathematics is currently taught, it can appear to have no obvious relationship to everyday life. When students are confused, they go off task instead of persisting with the task. Poor learning and negative emotional reactions can result.

The phenomenon of emotional reactions to task interruption in mathematics learning is worth more investigation because the phenomenon has ramifications for future learning that need to be understood. Not only are task interruptions the source of off-task behaviors but, as Kuhl (1986) and Heckhausen and Kuhl (1985) pointed out, task interruptions can have deleterious effects on the next task attempted. They said that later processing problems occur because the prior task was not completed and the individual cannot easily let go of it. It was initially cognitively encoded as an intention in a propositional framework, using verbs such as *can*, *wish*, *must*, and *will*. Tasks intended and not attempted, or not finished, become degenerated intentions. Because of the cognitive framework of these intentions, they are not dissipated. In Kuhl's view, this is because motivational and emotional states that result from these failed intentions are stored in the unconscious and operate simultaneously with conscious processing in attempting the next task; they take up processing capacity. This is another way that attention gets directed off task, and this is a way that so much emotionally charged information gets stored in the unconscious to interact with conscious processing.

It is this aspect of cognitive-emotional processing during learning that is particularly interesting. I am interested in the ways that emotionally laden stored information influences subsequent information processing. These memories can affect processing, even years later, depending on the current context of problem solving. In all cognitive and emotional processing, there appears to be a tradeoff between current processing of the actual event and processing that is dependent upon past experiences stored in long-term memory. When the topic is charged with emotion, there appears to be more possibility for distortion of current processing with information from the past, charged with self-esteem or lack thereof. Mandler makes a great deal of sense with his assertion that emotional reactions to mathematics learning have a lot to do with responses to difficult, confusing situations, ones similar to those in which students have had prior experience.

Although it is rarely applied to cognitive science or analyses of learning, the notion that information processing is dependent on past emotional experience has been discussed many times (for examples, Freud, 1915/1975; Jung, 1919/1960; Mandler, 1975; Tomkins, 1979; Zajonc, 1980). Zajonc, in fact, argues that the influence on current processing takes place so fast that emotion influences thought even before the cognitive reaction has occurred. Expectations that get built up over time sometimes unconsciously can determine current attitudes, likes or dislikes, and feelings. Tomkins, in his script theory, found that people processed information by reference to previously experienced "scenes" that get built up over time. In his work, he found that, at first, the actual occurrence of a scene (i.e., a mental model to organize incoming information) was most heavily influenced by the actual event, but when scenes were well established, the events were more influenced by the scenes; that is, previously experienced scenes determine subsequent processing. This is how otherwise rational, unemotional topics get processed emotionally. Attention is shared with information from the past, and when this information requires too much attention, there are not enough resources for the task, but because the information from the past is not being processed consciously, the influence from the past is hard to recognize and hard to change. It is understandable that a person learning a difficult mathematics problem may, following errors, become sidetracked into worries about self-concept and other "inefficient" topics. These topics are attitudes, beliefs, and feelings, which reside in the person's past for personal and cultural reasons. If students have expectations that they cannot perform well in mathematics learning, this may influence the actual course of learning.

Using this type of analysis of the cognitive-emotional processing interaction, we can understand our poor beleaguered mathematics student. If the information is confusing and errors have been causing task interruptions, then we can expect attention to be diverted from the task. A lot of this off-task information processing can be expected to concern the student instead of math. Prior experiences will have a lot of influence here, because how students generally handle frustration will play a big part in determining how they will react to the frustration of learning mathematics. The ways students handle frustration and errors and failure have a developmental precedence. There is a developmental precedence because many reactions to task processing and task failures have been internalized from experiences during early learning.

## Personality Development

Cognition and emotion are inseparable. Anything that happens cognitively has taken place in the context of the whole person, emotions included. Anything that happens emotionally is processed and stored cognitively. In fact, the very idea of a cognitive reaction implies emotional involvement. Although these processes may seem separate, they all operate simultaneously. The salience of individual differences in cognitive and emotional processing has origins in social conditioning. A particularly strong interaction comes from the developing individual and the parent. Later, when the child is in school, the interactions with the teacher become important. All learning takes place within the context of personality development, which happens within a relationship with another human being.

Cognitive growth and emotional growth occur simultaneously. This simultaneous growth occurs in the presence of and in interaction with other human beings, involving total dependence, at least in the beginning. It is in this meaningful context, involving acceptance or rejection at the deepest levels, that the individual develops. It is important to note that the individual's sense of competence, tolerance of frustration and failure, and need for completion and satisfaction, among other important personality variables, emerge and grow from this interaction. The reason these are such important variables in the learning environment is that they come about, or do not come about, because of emotional reactions to task interruption (Mandler, this volume, Chapter 1). Furthermore, there is a tendency among children, and human beings in general, to want to be correct, to maintain control for themselves, and to not be wrong or controlled (Deci, 1975). These are additional sources of emotional interaction between child and caretaker in the development of cognitive and emotional processing.

The interactions between child and caretaker during initial learning present many opportunities to learn how to behave in the learning environment, how to feel about oneself, how to handle frustration, and how to handle success. I believe that this aspect of cognitive and emotional interaction gives us an important clue about the interaction of cognition and emotion in the learning environment. In learning, many experiences from early childhood are internalized and no doubt remembered, perhaps unconsciously. These can affect what is being processed at a later time. When a difficult learning problem occurs, the memory structures present during past experiences have the tendency to be recalled, whether or not the individual is aware of the connection, and may operate in processing the current situation. For students who are not experiencing any difficulty in learning, these emotional interruptions are minor and infrequent, but for those who have to confront assaults on their egos, the reactions can be deleterious. Much depends on whether they have learned to tolerate errors, but this prior learning will not be apparent to the learner or the teacher in later situations. It is very automatic at this point.

The influence of previous experiences may be discerned in an experiment I conducted with military personnel playing a computer-based game (McDonald, 1987). This study was designed to investigate the interaction of cognitive processing and emotional processing during task performance. Both emotional reactions and objective performance criteria were monitored during regular game playing and game playing under stressful conditions. Data I obtained from subjects undergoing the stressful game experience showed that they talked a great deal about other non-task-related experiences. Very frequently, a subject's talk to the interviewer and the subject's self-talk (all tape-recorded) were not related to the task at all. These comments were related to experiences from the past—often experiences in which the subject had been under stress or in which the subject had failed or been perceived as a failure. Even those subjects who denied any experience of stress during the stressful game tended to talk about experiences that were seemingly non-task relevant. All subjects exhibited nonperformance behaviors, and, in some cases, their performance suffered significantly.

These results are compatible with the view advanced by Mandler (this volume, Chapter 1). If we were to observe the reactions in learning that stem from the cognitive-emotional interaction, we would not see pure, undistorted emotional reaction, because conscious emotional reactions can be distorted by the person experiencing them. Even though one's subjective experiential sense is that an emotional reaction is understandable, it is often difficult to understand all of the underlying dynamics of behavior (i.e., the unconscious emotional and motivational processing that accompanies behavior). Emotions are conditioned and processed cognitively, and the truth of an experience is difficult to disentangle— that is, unless you have the theoretical key to unlock the code. Many of us (including Mandler) think that this code lies in defense mechanisms.

Shevrin and Dickman (1980) said that there is a disjunction between people's behaviors and their conscious understanding of their behavior. Freud (1915/1975) said that the reason it is necessary to talk about the unconscious is because, in both healthy and sick persons, the "data of consciousness are exceedingly defective" (p. 99). The defense mechanisms that operate to maintain this defective thought range from denial to various forms of rationalization. Freud categorized the defense mechanisms according to what they do to the emotional reactions (e.g., fear of mathematics): (a) reversal into its opposite, or turning against the subject of fear ("I am not afraid of mathematics; I just find it boring." "Why do we need to learn this useless tool of the privileged?"); (b) repression ("No, I am not bothered by my math performance"); and (c) sublimation (spending a lot of energy to become the best athlete in town).[3]

Emotional reactions to learning are complex reactions, often involving processing influenced by past experience. Furthermore, the emotional-cognitive reaction is not necessarily easy to identify because of the distortions in people's thinking. However, understanding the cognitive-emotional processing interaction in this way helps us understand why people react emotionally in difficult situations. It provides a way to interpret emotion in the learning environment. Emotional reactions in learning may not be exactly what they appear to be on the surface; they may not be totally related to the present experience. However, the emotional reaction may be contributing to the way in which information is processed and stored for later use.

## Culturally Determined Values

In addition to emotional reactions internalized from early experiences with caregivers, other aspects of a person's self-concept and habitual response to learning are revealed when tasks are interrupted: culturally determined attributions for learning outcomes that were internalized. In fact, culturally determined values are worth exploring because differential learning in mathematics has become an issue. Although the cognitive analysis of reactions that are cultural is probably no different from that of the personal, I am discussing them separately because there is a different literature dealing with cultural beliefs and attitudes.

According to Mandler, a person is not going to have an emotional reaction to an experience that has no value or meaning. American education and reactions to mathematics merely reflect the tone of the culture. The lack of high achievement in mathematics by certain subcultures is particularly interesting, then, because the American culture has placed high value on science, mathematics, and technology since the Industrial Revolution. Why can mathematics have such a negative effect on some people? Perhaps the message communicated is that mathematics is so *important* that it is not for everybody. At any rate, mathematics learning is not an activity in which everyone participates. There is differential achievement in mathematics based on gender and race; that is why it is important for this discussion. These culturally determined values are influential in the cognitive analyses of ANS arousal in task interruption and give rise to attitudes about the self in the learning environment. This is part of the emotional reaction we observe in mathematics learning.

The importance of cultural influence on cognitive and emotional functioning lies in the beliefs and attitudes that are related to learning and how a person feels about himself or herself in the learning environment. These culturally driven expectations can look less affectively intense than the hotter emotions but can have a very intense influence on self-perception, self-regulatory strategies, and performance. Furthermore, the effects of these self-perceptions may not reach their full strength until years past formal education, even though they begin during the educational process (Dweck, 1986). These beliefs and attitudes take the form of attributions in the cognitive process. Through attributions of behavior, the self-concept is implicated in or protected from damaging emotional reactions. The motivational variables operating in the learning environment and the way in which they affect attributions of students is an interesting example of this cultural influence.

Dweck suggested that one reason for differential mathematics achievement is differential responsiveness between males and females to goals given in the learning environment. According to Dweck, performance goals (suitable for the measurement of actual performance) are used in education instead of learning goals (suitable for situations in which individuals are learning something new). The problem with using performance goals instead of learning goals in the learning environment is that it sets up a negative attributional process in some children. When competitive structures such as performance goals are used, children compare their ability with others, and effort is viewed negatively ("You must be smarter than me if you learned this in an hour"). If learning goals are used, children take pride in effort and the use of different strategies as opposed to simply relying on ability. It has been found that performance goals are associated with a *fixed* model of intelligence in the minds of children, whereas learning goals imply an *incremental* model of intelligence; therefore, children who think intelligence is fixed will choose performance goals, and children who think intelligence is incremental will do better with learning goals. Recent research has shown that performance goals work against the pursuit of challenge by requiring children's perceptions of their ability to be high and remain high. Students unfortunately will often choose tasks that conceal their ability or protect it from negative evaluation.

Unfortunately, there are differences between responsiveness of males and females to the types of goals that are set. Dweck said that females are more likely to suffer under performance goals, whereas males do better under competitive modes of performance goals. This is apparently because males attribute success to ability and failure to external causes such as luck or effort. As it turns out, the biggest differences appear among the bright students. Among students receiving As in mathematics, females do worse than males with performance goals. On task preference, these females prefer tasks on which they are assured of success, whereas the males prefer difficult tasks. Perhaps these attitudes would not have such a detrimental effect if different types of training were used or if different goals—ones that do not set up detrimental patterns of attribution—were used in learning.

The differential participation of students in this mathematical love affair has puzzled scientists and educators. For example, data on mathematics achievement between males and females show that males pull ahead of females in mathematics ability when they begin to learn algebra. The explanations for this have ranged from genetics to social conditioning. Benbow and Stanley (1980) suggested that the difference is due to the fact that the mathematical reasoning of boys is better developed than that of girls, which leads to boys taking more mathematics courses than girls. Fennema and Sherman (1977) believe it is the other way around. They said that females take fewer courses than males because of sociocultural factors and, thus, show less achievement. Parsons, Meece, Adler, and Kaczala (1982) suggested that the attributional style of females differs from that of males, especially when it comes to failure. Females learn helplessness in the learning situation, whereas males protect themselves from failure by attributing it to bad luck or to lack of effort. Betz and Hackett (1983, 1986) showed that females feel less capable in mathematics than do males, regardless of ability. Their studies showed that females feel capable in mathematics only when the problems are in stereotypically female areas. The females do not seem to generalize this ability to similar, more formal mathematics problems. By not taking the leap beyond these limited types of problems, females do not choose mathematically oriented professions, which pay well.

These considerations point to another way in which emotional reactions can occur in mathematics learning. They attest to the possibility that even relatively cool attitudes and beliefs can have extremely deleterious effects on learning performance. Although the beliefs or attitudes may represent cooler emotions, by the time they are influential in an individual student's information processing, they become hot. In the case of differential mathematics performance, females hold poor self-concepts with respect to their ability to survive failure and to perform ambiguous tasks, whereas males of similar ability do better on difficult-to-master tasks and in mathematics. Once an emotion, attitude, or belief is held that predisposes an individual to feel less than competent in the learning environment, the cognitive system is handicapped. As mentioned previously, these emotions, attitudes, and beliefs take up processing capacity and interfere with the energy available to put toward the task.

## A Leap to the Real World: Observations and Change

In this chapter, I have tried to explicate two different ways in which cognitive and emotional processes are involved in learning. One is through the individual representation of information that is tied to emotional concerns – the emotional reactions that affect moment-to-moment conscious processing. The other has to do with the sociocultural influence on individuals and the way that they see themselves or the information. This also provides a background of information processing against which new, incoming information is processed. In my view, both paths of influence come back to the self-concept and self-esteem, especially with respect to the task to be performed.[4] This self-esteem (as a representation of the self and its ability to complete the task or solve the problem) is unconscious, which makes it insidious, because all new information, even mathematics problems, is processed with this unconscious influence.

Convincing others that this connection is critical has not been easy, in part because of the difficulty of "seeing" the connection in a way that is influential in mathematics (or other) learning. I believe this is something that we can remedy empirically. Another reason for this difficulty is not so easy to remedy; it has to do with the difficulty of changing people's concepts of self, especially with respect to attitudes that have been culturally reinforced. Furthermore, in general, changing people's feelings requires a long process of retraining. (Consider the amount of psychotherapy required to change behavior and feelings.)

If we were to change students' feelings about mathematics and about themselves in learning, we would be talking about a major, long-term commitment to the issue. It would require programs of research and educational change across the grades and high school. I believe that this conference represents one of the first steps in achieving this goal. To change the general attitudes held about fear of mathematics would require change in top-level educators, parents, teachers, and students. It is a big order, but in order to change the continuing spiral of lowering student scores, I believe it is warranted. In my own work, I have concentrated on how emotions are implicated unconsciously, take up processing capacity, potentially interfere with task processing, and are then remembered with less than complete accuracy. Because of my interests, I have leaned toward certain kinds of research that I would include in my fantasy study of mathematics learning.

Whether the interest is in hot emotions, generated intrapsychically, or in cooler beliefs and attitudes, generated culturally, a method that converges on both processes in the mathematics learning environment is the study of the off-task topics that arise. I assume these topics would be related to the self-concept of and assessment of ability by students who do not handle mistakes and frustration well, who do not have a strong basis for learning mathematics, or who have been told that they probably will not be good in mathematics.

An example of this approach is illustrated by the study of computer-game playing under stress (McDonald, 1987) that was previously mentioned. In this study, subjects provided an ongoing conversation about their ability to perform a task; thus, collecting a tape-recording of all utterances and analyzing them with respect to the subject's actual performance proved to be beneficial. In order to make use of this technique, however, I assumed that what the subject said, on the surface, might not be related to the task or might appear to be completely nonemotional. If this is assumed, as it has been in my work, a content analysis of the types of statements made by subjects generally fall into a few categories. First, there are general statements about ability ("I am not good at this." "I usually do well at these types of tasks." "There I go, I don't know what I am doing again.") Second, there are statements about the task itself that indicate frustration or lack of frustration with the task ("This is a really interesting task." "This is a stupid task." "What is this supposed to mean?"). Third, there are comparisons made to other subjects performing the task that reveal evaluation concerns ("I know I am not as good as the other students." "How did John do on the task?"). Fourth, following errors or other task interruptions, subjects either continue on with the task or go off task, and this is of interest. If they go off task, they either express emotion directly or do so in some subverted way. For example, one of my subjects did not express anxiety directly; instead, he made jokes when he appeared to be anxious. During the first session (a nonstressful game), he made 16 jokes; during the second session (a stressful game), he made 45 jokes; and in the third session (a nonstressful game), he made 3 jokes.

In addition to my protocol analyses, I have employed individual-difference questionnaires. Clifford's Failure Tolerance Questionnaire (Clifford, 1988) and Sarason's Cognitive Interference Questionnaire (Sarason, 1975) can provide useful information for large studies. Any information tapping the students' preconceived attitudes toward the subject matter as well as their previous experience, as shown by achievement scores or a pretest, is always important. Tasks that have various levels of difficulty and that have points at which students can make choices about whether or not to pursue the learning experience are valuable.

Another way to study mathematics learning is to use cognitive task analyses and task decomposition techniques to get at a deeper understanding of where the student is making errors in the task and where the student is making cognitive errors in understanding. Decomposing the task may not provide clues about the emotional content, but it is the only way to get a good indication of what is going on cognitively. Marshall and Smith (1987) analyzed data from test performance by third graders and sixth graders by analyzing individual test items instead of total scores. By doing this, they were able to assess more complex skill acquisition and degradation. In addition, their analyses provided some interesting possibilities about where the cultural influence was making a difference in task performance. In other work, Marshall, Pribe, and Smith (1987) have studied mathematics learning in terms of the cognitive processes underlying problem solving. This appears to be a useful way in which to understand errors in mathematics learning.

In conclusion, mathematics learning need not be so difficult for students. All of us at the conference agreed upon that. The fact that it is often difficult for reasons not related to ability, particularly for some segments of the population, prompted the research in mathematics and emotion. Such research will guide instructional change. It is only by changing the environment to accommodate emotional and motivational reactions to mathematics that we can change students' willingness to withstand frustration, errors, and failure. Only then can we begin the job of making mathematics relevant to students' lives and easier to understand.

*Acknowledgments*.  The author wishes to thank Andrea Hattersley, James Crosswhite, and Ruth Kanfer for reading and commenting on an earlier draft of this chapter.

## Footnotes

[1]The views in this chapter are those of the author and do not necessarily reflect those of the Navy. Correspondence concerning this article should be addressed to the author at 4575 Orchard Avenue, San Diego, CA 92107.

[2]How do we categorize behavior in which one leaves the task (maybe or maybe not in frustration) to take a nap or to go fishing and, upon returning to the task, has the solution to the problem? Thanks to A. Hattersley and J. Crosswhite for bringing this consideration to my attention.

[3]Obviously, sublimation is a fairly constructive defense mechanism. It should be mentioned, however, that defense mechanisms do not necessarily operate in isolation. Sublimation probably operates in conjunction with repression.

[4]This is not, by any means, the only aspect of this interaction that is important to discuss; it is, however, my particular interest.

## References

Anderson, K.J., & Revelle, W. (1982). Impulsivity, caffeine, and proofreading: A test of the Easterbrook hypothesis. *Journal of Experimental Psychology: Human Perception and Performance, 8,* 614–624.

Beck, A.T. (1976). *Cognitive therapy and the emotional disorders.* New York: International Universities Press.

Benbow, C., & Stanley, J. (1980). Sex differences in mathematics ability: Fact or artifact? *Science, 210,* 1262–1264.

Betz, N.E., & Hackett, G. (1983). The relationship of mathematics self-efficacy expectations to selection of science-based college majors. *Journal of Vocational Behavior, 23,* 329–345.

Betz, N.E., & Hackett, G. (1986). Applications of self-efficacy theory to understanding career choice behavior. *Journal of Social and Clinical Psychology, 4,* 279–289.

Bower, G.H. (1981). Mood and memory. *American Psychologist, 36,* 129–148.

Clifford, M.M. (1984). Thoughts on a theory of constructive failure. *Educational Psychologist, 19,* 108–120.

Clifford, M.M. (1988). Failure tolerance and academic risk-taking in ten- to twelve-year-old students. *British Journal of Educational Psychology, 58,* 15–27.

Deci, E. (1975). *Intrinsic motivation.* New York: Plenum Press.

Dweck, C.S. (1986). Motivational processes affecting learning. *American Psychologist, 41,* 1040–1047.

Epstein, S. (1973). The self-concept revisited. *American Psychologist, 28,* 404–416.

Fennema, E., & Sherman, J. (1977). Sex-related differences in mathematics achievement, spatial visualization, and affective factors. *American Educational Research Journal, 14,* 51–71.

Freud, S. (1975). The unconscious. In J. Strachey (Ed. and Trans.), *The standard edition of the complete works of Sigmund Freud* (Vol. 14, pp. 159–215). London: Hogarth Press. (Original work published in 1915.)

Heckhausen, H., & Kuhl, J. (1985). From wishes to action: The dead ends and short cuts on the long way to action. In M. Frese & J. Sabini (Eds.), *Goal directed behavior: The concept of action in psychology* (pp. 133–159). Hillsdale, NJ: Lawrence Erlbaum Associates.

Humphreys, M.S., & Revelle, W. (1984). Personality, motivation, and performance: A theory of the relationship between individual differences and information processing. *Psychological Review, 91,* 153–184.

Izard, C.E. (1984). Emotion-cognition relationships and human development. In C.E. Izard, J. Kagan, & R.B. Zajonc (Eds.), *Emotions, cognition, and behavior* (pp. 17–37). New York: Cambridge University Press.

Jung, C.G. (1960). Instinct and the unconscious. In H. Read, M. Fordham, G. Adler (Eds.), & R.F.C. Hull (Trans.), *The structure and dynamics of the psyche: Collected works of C.G. Jung* (Vol. 8, pp. 129–138). New York: Pantheon Books. (Original work published in 1919.)

Kagan, J. (1984). The idea of emotion in human development. In C.E. Izard, J. Kagan, & R.B. Zajonc (Eds.), *Emotions, cognition, and behavior* (pp. 38–72). New York: Cambridge University Press.

Kanfer, R., & Ackerman, P. (1988). *Motivation and cognitive abilities: An integrative/aptitude-treatment interaction approach to skill acquisition.* Manuscript submitted for publication.

Kuhl, J. (1986). Motivation and information processing: A new look at decision making, dynamic change, and action control. In S. Sorrentino & E.T. Higgins (Eds.), *Handbook of motivation and cognition: Foundations of social behavior* (pp. 404–434). New York: Guilford Press.

Lazarus, R.S. (1984). On the primacy of cognition. *American Psychologist, 39,* 124–129.

Leon, M.R., & Revelle, W. (1985). The effects of anxiety on analogical reasoning: A test of three theoretical models. *Journal of Personality and Social Psychology, 49,* 1302–1315.

Leventhal, H., & Tomarken, A.J. (1986). Emotion: Today's problems. *Annual Review of Psychology, 37,* 565–610.

Mandler, G. (1975). *Mind and emotion.* New York: Wiley.

Mandler, G. (1984). *Mind and body: Psychology of emotion and stress.* New York: Norton.

Marcel, A.J. (1983). Conscious and unconscious perception: An approach to the relations between phenomenal experience and perceptual processes. *Cognitive Psychology, 15,* 238–300.

Marshall, S.P., & Smith, J.D. (1987). Sex differences in learning mathematics: A longitudinal study with item and error analyses. *Journal of Educational Psychology, 79*(4), 372–383.

Marshall, S.P., Pribe, C.A., & Smith, J.D. (1987). *Schema knowledge structures for representing and understanding arithmetic story problems* (Tech. Rep. Contract No. N00014-85-K-0661). Arlington, VA: Office of Naval Research.

McDonald, B.A. (1987). *The subtleties of cognitive and emotional expression.* Unpublished manuscript.

Parsons, J.E., Meece, J.L., Adler, T.F., & Kaczala, C.M. (1982). Sex differences in attributions and learned helplessness. *Sex Roles, 8*(4), 421–432.

Revelle, W. (1987). Personality and motivation: Sources of inefficiency in cognitive performance. *Journal of Research in Personality, 21,* 436–452.

Sarason, I.G. (1975). Anxiety and self-preoccupation. In I.G. Sarason & C.D. Spielberger (Eds.), *Stress and anxiety* (Vol. 2, pp. 27–44). Washington, DC: Hemisphere.

Shevrin, H., & Dickman, S. (1980). The psychological unconscious: A necessary assumption for all psychological theory? *American Psychologist, 35,* 677–688.

Tomkins, S.S. (1979). Script theory: Differential magnification of affects. In H.E. Howe, Jr., & R.A. Dienstbier (Eds.), *1978 Nebraska symposium on motivation* (pp. 201–236). Lincoln: University of Nebraska Press.

Zajonc, R.B. (1980). Feeling and thinking: Preferences need no inferences. *American Psychologist, 35,* 151–185.

# Part V   Looking Back

# 16
# Affect and Learning: Reflections and Prospects

GEORGE MANDLER

My comments on our conference reflect the usual amalgamation of one's initial presuppositions and prejudices with the new learning that has taken place. Having come to the mathematical learning field as a novice, armed only with some theoretical notions about the sources of emotional reactions, I now understand better some of the pervasive problems of the affective side of mathematical problem solving. My remarks in this section focus on what I have understood to be issues that deserve more extensive exploration. I start with a discussion of some definitional problems, and then move on to a further exploration of values and their effect. I then want to ask some questions about the origins of affect in the mathematical domain. This is followed by an outline of possible directions of research, including some work that, as a result of our participation in these meetings, Carmen Overson and I plan to pursue.

## Definitions

I sympathize with Laurie Hart's concern with definitions and the lack of communication that ill-defined concepts engender (Hart, this volume, Chapter 3). But, absent some theoretical agreement, a search for consensus may well be futile. The social sciences are well served by common sense and natural language concepts in the early stages of exploring a field. During the development of a field of concern, it may be best to rely on our understanding of common meanings and natural language. Much of our talk and our concepts are metaphoric, and we understand what we say; however, when we start to make significant progress, when we reach some degree of precision, we need theory to guide our concepts, and theory must and will inform our definitions. Neologisms may then have their place, and theoretical legislation for their use will be appropriate. The good theory can well say with Humpty Dumpty: "When I use a word, it means just what I choose it to mean – neither more nor less." However, the history of the social sciences is strewn with abandoned concepts and terms that have failed to heed a corollary that Humpty Dumpty never told us about: Once you choose a word to mean something (exactly), then you have to start convincing other people to use it the same way; otherwise, monologues will never be replaced by dialogue and consensus.

There is one concept that we need to be more careful with, even before we reach any theoretical agreement as to its "proper" use; that concept is affect. At the present time, affect is used to cover a multitude of sins and virtues, and its application ranges from the preference for chocolates to the terror of possible nuclear war. The dictionary is not much help; it defines the noun *affect* as, among other things, a "disposition of body or mind; affection, love; the emotion that lies behind action; pleasantness or unpleasantness of, or complex of ideas involved in, an emotional state" (*Chambers Twentieth Century Dictionary*, 1977). In my introductory chapter, I implied an indiscriminate use of the concept, but I now want to withdraw that suggestion. In keeping with most (although not all) common uses, I would now prefer not to use the term *affect* synonymously with the term *emotion*, which implies the hot passions, whether fear or joy, anger or love. Affect should be restricted to the cognitive aspects of emotions, to the values, attitudes, beliefs, and ideas that instruct the quality of emotions, but not to the full emotions themselves. Only then can we make a fairly clean distinction between affect and emotion. For my purposes, this distinction is important because affects are then coextensive with what I have sometimes called *evaluative cognitions*. Emotions, on the other hand, require some kind of (presumably autonomic) arousal.

## Values

If, as I claim, the nature of our emotions is a function of the values operating and invoked at the time the "emotion" occurs, then the study of values becomes a central issue if we wish to change the emotional climate in which mathematical learning and problem solving occur. Many of these values are cultural and social ones; they come, in a real sense, with one's mother's milk. Parents, and later peers and teachers, are the primary conveyors of cultural values, of the positive and negative evaluations we imposed on our world. Despite much talk about the rebelliousness of children and adolescents, in particular, it is the case that the values of children are more likely to be similar to, rather than different from, those of their parents.

We need to be careful, however, about invoking the cultural transmission of values. Such a concept does not imply that a value or range of values is acquired and then simply and straightforwardly applied in all situations and at all times. Values and their applications are often situation specific. Consider the fact that most of us apply values of honesty and truthfulness quite variably; we are willing to lie to a bureaucrat who has kept us needlessly waiting but not to our friends and loved ones (but even that distinction has fuzzy edges). Similarly, it need not be the case that children apply their parents' values indiscriminately. Consider the observation that children's aversion of mathematics does not seem to appear until 1 or 2 years of schooling have gone by. It is easy to assign the cause of this change to the teachers; after all, the kids are quite willing to entertain mathematical thinking in first grade but are often skittish about the subject by the third grade. It may take that long for parental values about mathematics to emerge, possibly because the rather interesting number games are not like that thing a parent has described as being difficult or irrelevant or not suited to their temperament. Parental values emerge when the situation is perceived as appropriate.

I do not wish to imply that any of this is a question of placing blame on parents or on teachers or peers. Rather, I would argue that cultural attitudes toward a topic are transmitted and reinforced by all of these groups; they are the carriers of culture. For example, Americans do not like certain foods (e.g., frogs' legs) without having tasted them, and similar prejudices about foods (and appearance, music, and social conventions, etc.) can be found in all cultures. What I am concerned with are cultural attitudes that are pervasive across different groups and subcultures. One such cultural characteristic is our negative and, at best, cavalier attitude toward mathematics learning and use. Someone once mentioned that our children ought to do better in mathematics so that we can "compete with the Japanese." The easiest way to do that would be to incorporate Japanese values into our culture, but that is impossible because cultures are not easily manipulable. I would like to see more research on this question of "mathematical values." What are adults' and children's attitudes toward mathematics, and how do these attitudes vary across classes and other subgroups in our culture? When do they emerge, and how intensely are they held? For the time being, one can observe that we rarely see math heroes on that great cultural transmitter, the television screen. We cannot forget that other transmitter of values, the actions and inactions of our masters. Governmental policy, when implemented, also tells us what is good and bad, and it may be most influential when it is ambivalent, when it sends out positive messages that are countermanded by negative actions. For example, the Congress and the President proclaimed 1987 as the "Year of the Reader," an action that was aimed at restoring "the act of reading to a place of pre-eminence in our personal life and in the life of our nation." Yet the administration's budget for fiscal 1988 proposed to reduce aid to public libraries by more than 25%. What kind of ambivalent, or even negative, values do youngsters adopt when they are urged to read more, and then find their local library closed?

It should be understood that my concern with values is at heart a concern with emotions. It is the values that determine the quality of an emotion when we experience autonomic arousal. Therefore, it is the values of mathematics that the learner brings to the situation that will color the emotions that are experienced. If errors, mistakes, and interruptions occur, which is inevitable, the potential mathematicians will experience the resulting arousal as negative if attitudes toward the subject are generally negative. If, however, mathematics is seen as challenging and positively valued, some of these occasions may be experienced as positively arousing and as steps toward mastery and success. An erroneous strategy may be seen as a challenge to find a better one rather than as a signal of failure and incompetence.

## Origins and Manipulations of Affect

I have traveled rather far afield from an examination of cognition and affect. If it is the case, however, that cultural values are one of the obstacles to a mathematically literate society, we need to face that issue. What can we do? Obviously, we cannot change the culture; we cannot produce TV programs about heroes of the calculus, winners of the Field prize, and emulators of Euler. We can try to influence parents' attitudes, but first we need to know more about them. It may be easier to convince teachers that little children are curiosity machines whose delight in mathematics can be nurtured as long as the subject is presented as positive and exciting — not full of pitfalls and right and wrong answers. We can also take to heart one of the themes of our discussion — namely, that children need not necessarily enjoy mathematics in order to do it well. It can be seen as something worthwhile doing for a multitude of positive goals, even including that of "competing with the Japanese." One possible avenue has been indicated by Silver and Metzger (this volume, Chapter 5): the exploitation of the aesthetic values of mathematics. If children were to come to grips with the beauty of mathematics, we would have gone a long way toward making it not only palatable, but desirable.

At the cultural level, we need more information about the sources of mathematics' negative reputation. Is the frequent self-label "mathematical idiot," a defense mechanism? Does this label make it possible to avoid an unpleasant topic? If so, such self-deprecation may well hide, rather than display, emotion. Furthermore, these and similar defensive maneuvers may be adopted by children. Does the child who is "turned off" by mathematics simply play "idiot" in another way?

Longitudinal studies are long-term investments, and our current culture prefers short-term investments and immediate goals; however, I believe that the longitudinal study of children's experience with mathematics may have important payoffs in telling us how the "curiosity machine" turns into a "mathematical idiot." When in a child's life of mathematics do the first signs of aversion to mathematics appear? How are these signs first expressed in the learning situation? What does the child notice about parental attitudes and beliefs? How is their effect shown? When? And while we are on the topic of long-term and short-term goals, how do we inculcate a tolerance for error? Such a tolerance would ameliorate the degree of arousal and the intensity of emotions, and seeing an error as just another step toward mastery would, of course, make it less aversive and, at times, even positively valued.

## Directions

In the preceding sections, I have indicated several categories of needed information. I now turn to some more specific suggestions for future directions.

## Protocol Analyses

Cognitive psychology has greatly benefited from the analyses of think aloud protocols (see Newell & Simon, 1972). We have learned that people can be relatively easily induced to speak their current thoughts, hypotheses, and strategies in the process of solving problems. The analyses of these externalized thoughts have provided important indicators to the underlying cognitive processes that generate intelligent behavior and efficient problem solution.

It is generally assumed that the emotional events that intrude on effective mathematical problem solving arise out of an interaction of ongoing problem-solving activities and affective events such as aversions, hopes, and expectations. We know relatively little, however, about the course of that interaction, and about the sequence of intrapsychic events that produces emotion in the mathematics arena. I would suggest that careful analyses of ongoing thought processes during mathematical problem solving could provide one avenue to a better understanding of the interplay between affect and cognition. There is every reason to expect people to be able to verbalize their ongoing feelings and emotional states, as well as their conscious "cognitive" processes, as they solve problems and do mathematics. What is the sequence of internal (conscious) events that leads to a failure experience? When and how does a learner react fearfully (or hopefully) to a specific problem or kind of problem? What kind of strategic or defensive mechanisms are employed to deal with actual or potentially emotional problems or situations in the course of mathematical thinking? Is there a general affective mode that characterizes some students, or are these affective events situation specific? How does remedial intervention affect subsequent problem solving? Does it differ in specifiable ways from pre-intervention thinking? These are the kinds of questions that could be fruitfully explored by the use of protocol analyses.

## Is It Useful to Think of Math Anxiety as a Quasi-Neurosis?

I was impressed by Alba Thompson's presentation of the case of Marcus, who, in the midst of obvious mathematical blocks and failures, seemed to show no or little emotional reactions (Thompson & Thompson, this volume, Chapter 11). To what extent is such a reaction similar to the bland accounts of neurotic patients who deny or avoid their real-life difficulties? As I indicated before, the self-description of "mathematical idiots" by many adults may have a similar defensive purpose. If I am a mathematical idiot, I do not have to face doing mathematics; I can avoid this type of emotional episode altogether. Does this differ from the therapeutic clients who avoid social interactions and thereby escape the consequences of their painful inability to develop social relationships? Did we see in Marcus the beginnings of the "mathematical idiot" who, at age 10 years, is turned off by mathematics and engages in irrelevant, extraneous behavior to avoid the unpleasant emotions that arise when actually engaging the task? Will these childhood mechanisms turn into a full-blown avoidance of the problem by ascribing the denial and avoidance of mathematics to some "innate" characteristic of mathematical idiocy?

If such an approach has any value, then it might be interesting to explore the mechanisms of denial and displacement that we apparently see in children and adults. One can speculate about how an experienced psychological counselor might approach Marcus' attitude and how these symptoms could be understood and ameliorated. I am not suggesting that the problem of affect in mathematical problem solving can or should be dealt with in the context of individual counseling sessions, but an analysis of frequently seen mechanisms and symptoms might point us in the direction of more general ameliorative measures.

Related to this way of asking our questions is the definition of the "good teacher." We heard a number of different descriptions of successful teachers who could deal with and avoid serious emotional interactions in mathematical problem solving. Carmen Overson suggested that the discussion sounded at times much like the equally groping searches for the "good therapist." If, as I believe, the parallel is real, then we are looking for teachers who communicate acceptance to their students (no matter how disappointing their performance) and who elicit trust by demonstrating their involvement in the students' problems. Teachers who will understand the relative salience of different attitudes and values, who will know when to intervene and when to abstain from doing so, should provide the emotional and intellectual support that the affectively handicapped student of mathematics needs.

## Some Projected Research

I indicated before that one result of these conferences has been the initiation of new directions in our laboratory to be undertaken by Carmen Overson and me. One focus of our investigation will be the physiological consequences of errors (misreadings, mistakes, etc.) during mathematical problem solving. The research will involve an individual solving problems that will generate error-prone situations, while physiological responses (heart rate and galvanic skin response) are measured. We also shall obtain on-line ratings from subjects of the intensity of their subjective affective reactions while performing selected tasks.

Our theoretical position, buttressed by available evidence, is that much of the interfering effects of emotional reaction derive from the attention-demanding character of autonomic arousal and subsequent subjective phenomena. Looking at emotion from that point of view, three of the more important issues about the relationship between emotion and problem solving are as follows:

1. What events in mathematical problem solving result in autonomic arousal?
2. What effects does arousal have on subsequent mathematical problem-solving behavior?
3. How does autonomic arousal interact with various aspects of mathematical problem solving in general?

Because autonomic arousal primarily determines the intensity of an affective reaction, we shall, of course, at all times also be concerned with the quality, or the evaluative aspect, of these responses.

One of the most obvious issues to be addressed is which events in mathematical problem solving bring about autonomic arousal. Arousal is hypothesized to occur whenever a person encounters an unexpected (interrupting) event, specifically errors and mistakes. However, unexpected success as well as unexpected failure on a particular problem should evoke arousal. Likewise, being "stuck" on a problem or subroutine will result in arousal. We need to determine what effect expected successes and failures have on autonomic arousal, and how this differs from the unexpected case.

All these questions are "within" problems – that is, they are about the effects of unexpected occurrences while solving a particular problem. There is a corresponding set of questions about occurrences "between" or "among" problems. Suppose a person works several problems in a row that are all quite similar. What will happen when she or he is faced with a new type of problem? What if that problem looks like the previous ones but is unexpectedly easy or difficult? What if the new problem does not look like the others, but in fact has a similar structure? Other types of unexpected events that could be investigated are those that are unrelated to the problem at hand; for example, how might sudden changes in room temperature, lighting, and noise levels affect arousal, affect, and concurrent problem solving?

Along with determining which aspects of doing mathematics result in arousal, another set of questions involves the effects such arousal has on subsequent problem solving. As discussed earlier, one effect of arousal is a focusing of attention on the salient aspects of the situation. Saliency, of course, is defined by the individual; what a person views as important about a particular problem-solving situation may not coincide with the researcher's, teacher's, or mathematician's view of things. Theoretically, then, it is possible that arousal in one person may lead him or her to focus on the aspects of a problem that will lead to a correct solution, whereas someone else will focus on extraneous information. Arousal, in this sense, is not in and of itself either good or bad; it merely magnifies what the person brings to the problem. In the single-problem case, it would be useful to examine what effects arousal and affect have on problem-solving behavior. For instance, are subjects more or less likely to go back and check their work when they are aroused than when not? After finding themselves stuck on a problem, will aroused subjects try again by using a new technique, try the same method again, or quit? Are aroused subjects more or less likely to recognize false starts and fruitless methods than nonaroused subjects? Similarly, when working on a set of problems, will aroused subjects emphasize even more the similarities among problems (which may or may not be beneficial)? Do subjects work harder when aroused, whether or not they are pursuing a correct solution? Does arousal over a somewhat longer term facilitate or hinder learning? Are effects of arousal due to unrelated interruptions (e.g., change in lighting or noise) in any way different from those due to pleasant or unpleasant interruptions that *are* related directly to problem-solving activity? Such distractions might have negative effects for someone who is not otherwise having difficulties with the problem at hand. Could they ever be beneficial? Answers to these questions might have important consequences for actual teaching situations.

The third issue concerns the interaction of autonomic arousal with various aspects of problem solving. Previous researchers (Briars, 1983; Mayer, 1982; Schoenfeld, 1985) have proposed slightly different theoretical frameworks for understanding mathematical problem solving. Mayer discusses the importance of four types of knowledge: linguistic, schematic, algorithmic, and strategic. Schoenfeld focuses on resources, heuristics, control, and belief systems in doing mathematics. Briars describes mathematical ability as basic information-processing skills, content knowledge, and metacognition. Our central question at this stage is not which, if any, of these theories is the best way to understand what people are doing as they solve math problems. However, the theories do suggest cognitive factors that might interact with autonomic arousal. It would be important to know how arousal affects and is affected by these factors. Autonomic arousal may interact with various low-level cognitive functions; for example, even though a subject has relevant background knowledge represented in memory, she might retrieve it more or less easily when aroused. Arousal might also affect how well subjects read and understand a problem, and represent it both to themselves and on paper. Some subjects might be able to use heuristics and carry out computation smoothly when not aroused, but these abilities could also be affected by arousal.

At this point, we are not sure which of these possible avenues we shall explore in detail. We have been successful in investigating the effects of expectations and discrepancies and their relation to autonomic (sympathetic) arousal in previous studies, and we have every expectation that this line of research will contribute to some extent to our understanding of affect and mathematical problem solving.

*Acknowledgments.* Preparation of these remarks was supported by grants from the Spencer Foundation and the National Science Foundation. I am grateful to Carmen Overson for important contributions to this paper.

## *References*

Briars, D.J. (1983). An information-processing analysis of mathematical ability. In R.F. Dillon & R.R. Schnieck (Eds). *Individual differences in cognition* (Vol. 1, pp. 181–204). Orlando, FL: Academic Press.

Mayer, R.E. (1982). The psychology of mathematical problem solving. In F.K. Lester & J. Garofalo (Eds)., *Mathematical problem solving: Issues in research* (pp. 1–13). Philadelphia: Franklin Institute Press.

Newell, A., & Simon, H.A. (1972). *Human problem solving.* Englewood Cliffs, NJ: Prentice Hall.

Schoenfeld, A.J. (1985). *Mathematical problem solving.* Orlando, FL: Academic Press.

# 17
# Beliefs, Attitudes, and Emotions: New Views of Affect in Mathematics Education

Douglas B. McLeod

All of the chapters in this book discuss research and theory related to the role that affect plays in mathematical problem solving. This chapter begins with a reanalysis of the affective domain, describing affect in terms of beliefs, attitudes, and emotions. The discussion of these three topics indicates the broad impact of affective factors in mathematics learning and relates beliefs, attitudes, and emotions to key chapters in this book. The chapter continues with a discussion of several broad themes that recur with some regularity in this book, including the central role of affect in problem solving, the need to integrate research on cognition and affect, and the importance of the social context in the study of affective factors in mathematics learning. The chapter concludes with comments on some methodological issues and their implications for future research on affective factors in the learning and teaching of mathematics.

## Affective Factors in Mathematics Learning

The theoretical analyses of Mandler (1984), the discussion of the affective domain by Hart (this volume, Chapter 3), and observations of mathematics classrooms (e.g., see Grouws & Cramer, this volume, Chapter 10 or Lester, Garofalo, & Kroll, this volume, Chapter 6) suggest that beliefs, attitudes, and emotions are important factors in research on the affective domain in mathematics education. When this book was in the planning stages, the major emphasis was on the emotional aspects of mathematical problem solving. At that time, my view was that the emotions were the most underrepresented of the various aspects of affect, so it seemed appropriate to place more emphasis on emotional factors. The research reported in this book, however, approaches the affective domain in a more comprehensive way. This section of the chapter reflects that broader view of affect and outlines how research on beliefs, attitudes, and emotions could provide a more complete approach to understanding affective factors in mathematical problem solving.

The term *affective domain* is used here to refer to a wide range of feelings and moods that are generally regarded as something different from pure cognition.

Beliefs, attitudes, and emotions are terms that express the range of affect involved in mathematical problem solving. These terms vary from cold to hot in the level of intensity of the feelings that they represent. They also vary in their stability; beliefs and attitudes are generally thought to be relatively stable and resistent to change, but emotional responses to mathematics may change rapidly. For example, students who say that they dislike mathematics one day are likely to express the same attitude the next day; however, a student who is frustrated and upset when working on a nonroutine problem may express strong positive emotions just a few minutes later when the problem is solved.

Also, beliefs, attitudes, and emotions differ in the ways that cognition is involved in the affective response. Although it is not possible to separate student responses into affective and cognitive categories, some of these terms have a larger cognitive component than others. For example, beliefs are mainly cognitive in nature; they are built up rather slowly over a relatively long period of time. Emotional responses, however, have a much stronger affective component, and their rise time can be very short. The terms *beliefs*, *attitudes*, and *emotions* then, are listed in order of increasing affective involvement, decreasing cognitive involvement, increasing intensity, and decreasing stability.

## Beliefs

One of the benefits of the cognitive revolution in psychology is that it has brought together people from different disciplines who all proceed on the basis of the same (or at least similar) fundamental assumptions about human behavior. In the study of beliefs, mathematics education can benefit from the experience, theories, and methodologies of cognitive anthropologists and others who study the social context in which we live (Cole & Griffin, 1987; D'Andrade, 1981; Lave, Murtaugh, & de la Rocha, 1984).

Two major categories of beliefs seem to have an influence on mathematics learners. First, students develop a variety of beliefs about mathematics as a discipline. These beliefs generally involve very little affect, but they form an important part of the context in which affect develops. A second category of beliefs deals with students' (and teachers') beliefs about themselves and their relationship to mathematics. This category has a stronger affective component and includes beliefs that are related to confidence, self-concept, and causal attributions of success or failure. There is a substantial amount of research that deals with both kinds of beliefs.

### BELIEFS ABOUT MATHEMATICS

Research on student beliefs about mathematics has received considerable attention over recent years. The National Assessment of Educational Progress (Brown et al., 1988) has included items related to beliefs about mathematics for some time. The most recent assessment results indicate that students believe that mathematics is important, difficult, and based on rules. These beliefs about mathematics, although not emotional in themselves, certainly would tend to generate more intense reactions to mathematical tasks than beliefs that mathematics is unimportant, easy, and based on logical reasoning.

Research on beliefs has been highlighted by the results of research on problem solving. As Schoenfeld (1985) and Silver (1985) have pointed out, students' beliefs about mathematics may weaken their ability to solve nonroutine problems. If students believe that mathematical problems should always be completed in 5 minutes or less, they may be unwilling to persist in trying to solve problems that may take substantially longer. Nevertheless, this kind of belief has been generated out of the typical classroom context in which students see mathematics. There is nothing wrong with the students' mechanism for developing beliefs about mathematics (D'Andrade, 1981); what needs to be changed is the curriculum (and beyond that, the culture) that generates such beliefs.

Another important area of research on beliefs comes mainly out of the work on gender-related differences in mathematics education. Most of the data have come from studies that used the Fennema and Sherman (1976) scales, especially the scale on the perceived usefulness of mathematics. Fennema (this volume, Chapter 14), in summarizing this research, notes that, in general, males report higher perceived usefulness of mathematics than females. Other scales (e.g., mathematics as a male domain) also deal with beliefs about mathematics. These kinds of beliefs are important both for gender-related differences in mathematics achievement and for the related differences between females and males in affective responses to mathematics.

## Beliefs About Self

Research on self-concept, confidence, and causal attributions related to mathematics tends to focus on beliefs about the self. These beliefs about self are closely related to notions of metacognition and self-awareness. Again, in mathematics education, the research on gender-related differences has taken the lead in this area.

Research on self-concept and confidence in learning mathematics indicates that there are substantial differences between males and females on this dimension. Reyes (1984) and Meyer and Fennema (1988) summarize the relevant literature. In general, males tend to be more confident than females, even when females may have better reason, based on their performance, to feel confident. The influence of confidence on mathematical performance, especially in the area of nonroutine problem solving, seems relatively direct.

Another set of beliefs about self has been investigated quite thoroughly under the rubric of causal attributions of success and failure. Although there are several antecedents of this work and many different applications of the ideas, the central themes are well explicated in a recent reformulation of the theory by Weiner (1986). The three main dimensions of the theory deal with the locus (internal or external), the stability (e.g., ability versus effort), and the controllability of the causal agent. For example, a student who fails to solve a mathematics problem could say that the problem was too hard, a cause that is external, stable, and uncontrollable by the student. A student who succeeds in solving a problem might attribute that success to effort, a cause that is internal, unstable, and controllable.

The nature of the attributions of female and male students has been an important theme in recent research in mathematics education, and the results of this research provide some of the most consistent data in the literature on the affective domain. For example, males are more likely than females to attribute their success in mathematics to ability, and females are more likely than males to attribute their failures to lack of ability. In addition, females tend to attribute their successes to extra effort more than males do, and males tend to attribute their failures to lack of effort more than females do. The resulting differences in participation in mathematically related careers appear to reflect these gender-related differences in attributions (Fennema, this volume, Chapter 14; Fennema & Peterson, 1985; Meyer & Fennema, 1988; Reyes, 1984).

## TEACHERS' BELIEFS

So far our discussion has concentrated on students' beliefs about mathematics and about themselves, but the corresponding set of beliefs that teachers hold about mathematics, mathematics teaching, and themselves are also important to the study of affect in mathematics education. There have been a number of important studies of teachers' beliefs about mathematics (e.g., Thompson, 1984), and current recommendations for a research agenda on mathematics teaching suggest that more work should be done in this area (Cooney, Grouws, & Jones, 1988). The studies of teaching in this book (e.g., J. Sowder, this volume, Chapter 12) are important steps in implementing that research agenda. There are other studies of teachers' beliefs and attributions that also help to further our understanding of affective factors in classroom instruction; most of these studies, however, do not deal specifically with mathematics teaching. See Peterson and Barger (1985) and Prawat, Byers, and Anderson (1983) for important examples of this work.

In summary, research on beliefs and their influence on students and teachers has been an important theme in investigations of learning and instruction in mathematics. Some of this research is directly connected with affective issues (e.g., confidence), but much of it is not. Because beliefs provide an important part of the context within which attitudinal and emotional responses to mathematics develop (Mandler, this volume, Chapter 1), we need to establish stronger connections between research on beliefs and research on other aspects of the affective domain. Moreover, beliefs about mathematics play a central role in problem solving by experts (Silver & Metzger, this volume, Chapter 5) as well as novices (Lester, Garofalo, & Kroll, this volume, Chapter 6), and by teachers as well as students (J. Sowder, this volume, Chapter 12). Clearly, research on beliefs is crucial to the success of research on other aspects of affect and mathematical problem solving.

## *Attitudes*

Research on attitudes towards mathematics has a relatively long history compared to other areas in the affective domain. For recent reviews and analyses, see Haladyna, Shaughnessy, and Shaughnessy (1983), Kulm (1980), Leder (1987), and Reyes (1984). In this chapter, the term attitude is reserved for affective responses that involve positive or negative feelings of moderate intensity and reasonable stability. Examples of attitudes toward mathematics would include liking geometry, disliking story problems, being curious about topology, and being bored by algebra. As Leder (1987) and others have noted, attitudes toward mathematics are not a unidimensional factor; there are many different kinds of mathematics, and a variety of different feelings about each type of mathematics.

Attitudes toward mathematics appear to develop in two different ways. First, attitudes may result from the automatizing of a repeated emotional reaction to mathematics; for example, if a student has repeated negative experiences with geometric proofs, the emotional impact will usually lessen in intensity over time. Eventually, the emotional reaction to geometric proof will become more automatic, there will be less physiological arousal, and the response will become a stable one that can probably be measured through use of a questionnaire. A second source of attitudes is the assignment of an already existing attitude to a new but related task. A student who has a negative attitude toward geometric proof may attach that same attitude to proofs in algebra. To phrase this process in cognitive terminology, the attitude from one schema is attached to a second, related schema. For a more detailed discussion of a cognitive approach to the formation of attitudes, see Abelson (1976) and Marshall (this volume, Chapter 4).

If we think of attitudes as the end result of emotional reactions that have become automatized, then we might predict that there is an attitude that corresponds to every emotion. Students who are fearful in certain mathematical settings can eventually become chronically anxious. If a student regularly has positive experiences with nonroutine mathematical problems, an attitude of curiosity and enthusiasm regarding problem solving could develop.

Research on attitude toward mathematics has generally proceeded in isolation from more contemporary, cognitive approaches to research on learning (Leder, 1987). The work of Marshall, L. Sowder, and Kaput in this book should help to encourage other cognitive researchers to include considerations of attitude in their studies of learning and problem solving.

## *Emotions*

The emotional reactions of students have not been a major factor in research on affect in mathematics education. In part, this lack of attention to emotion is probably due to the fact that research on affective issues has primarily looked for factors that are stable and can be measured by questionnaire. A number of studies, however, have looked at the processes involved in learning mathematics, and these studies have sometimes paid attention to the emotions. In this section, we review briefly a few of these studies.

Reports of strong emotional reactions to mathematics have not appeared in the research literature very often. An important exception is the work of Buxton (1981). His research deals with adults who report their emotional reaction to mathematical tasks as "panic." Their reports of panic are accompanied by a high degree of physiological arousal; this arousal is so difficult to control that they find it disrupts their ability to concentrate on the task. The emotional reaction is described as fear, anxiety, and embarrassment, as well as panic. Buxton interprets these data in terms of Skemp's (1979) views of the affective domain and suggests a number of strategies to change students' beliefs in order to reduce the intensity of the emotional response.

A number of researchers who focus mainly on cognition have also noted the influence of emotions on cognitive processes in mathematics. Wagner, Rachlin, and Jensen (1984) report how algebra students who were stuck on a problem would sometimes get upset and grope wildly for any response, no matter how irrational, that would get them past the blockage. On a more positive note, Brown and Walter (1983) discuss how making conjectures can be a source of great joy to mathematics students. Similarly, von Glasersfeld (1987) notes the powerful positive emotions that often go along with the construction of new ideas or the cognitive reorganization of old ideas. Lawler (1981) also documents the positive emotional responses that accompany that moment of insight when a child first sees the connections between two important ideas.

Although comments about emotion do appear in the research literature from time to time, it is fairly unusual for research on mathematics education to include measures of the physiological changes that accompany the emotions (Mandler, 1984, this volume, Chapter 1). An exception is a recent study by Gentry and Underhill (1987) in which they used physical measurements of muscle tension as well as paper-and-pencil measures of anxiety toward mathematics. As one might expect, there was little correlation between the two measures, suggesting that traditional measures of anxiety may be quite different from the emotional responses that influence students in the classroom.

In summary, research on emotional responses to mathematics has been with us for some time, but it has never played a prominent part in research on the affective domain in mathematics. A major problem has been the lack of a theoretical framework within which to interpret the role of the emotions in the learning of mathematics. This book, with its emphasis on the application of Mandler's (1984) theory to mathematical problem solving, continues the process of building up that theoretical framework. Silver and Metzger (this volume, Chapter 5), for example, note how positive emotions are an important part of the aesthetic forces that influence expert problem solvers, and Cobb, Yackel, and Wood (this volume, Chapter 9) report how a teacher helped students deal with their emotional reactions to mathematics in a second-grade classroom. The available data from a variety of sources suggest that emotion plays an important role in mathematical problem solving and that research can make progress in understanding that role.

Describing the affective domain in terms of beliefs, attitudes, and emotions, as suggested by Hart (this volume, Chapter 3), provides a useful structure for analyzing affective factors in mathematical problem solving. This analysis needs to proceed further in order to clarify how various other lines of research (e.g., self-concept, learned helplessness) can be more fully integrated into research on mathematics learning.

## Recurring Themes and Promising Directions

Silver (1985), in an influential analysis of mathematical problem solving, suggested a number of areas in which more research was needed. He identified affect as one of the more important underrepresented themes in this research area. In this section, we review briefly some of the same issues that Silver (1985) discussed, but in this case we review them in terms of their relationship to affective issues and in the context of several years' progress in research on problem solving.

### *The Central Role of Affect in Problem Solving*

The initial hypothesis of this project was that affect played an important role in problem solving, and that researchers who observed carefully would see the evidence of affect in both students and teachers. That hypothesis has been confirmed. The chapters in this book provide a wealth of data to document the ways in which beliefs, attitudes, and emotions are related to the performance of both students and teachers in mathematics classrooms.

The recent publication of national standards for the mathematics curriculum (Commission on Standards for School Mathematics, 1987) reaffirms the centrality of affective issues to mathematics learning. Two of the major goals stated in this document deal with helping students understand the value of mathematics and developing student confidence. In addition, the standard on the development of a mathematical disposition emphasizes the importance of student confidence, interest, perseverance, and curiosity. To this list we might add that the role of the emotions needs attention.

The discussions at our conferences often focused on the central role of affective factors in classrooms as well as in research. In these discussions, Joe Crosswhite, past President of the National Council of Teachers of Mathematics, pointed out that teachers have always paid attention to affect in the classroom, even when researchers did not. However, the new emphasis on problem solving in mathematics classrooms presents teachers with new challenges. When students work on nonroutine problems, their affective responses are more intense; we see more evidence of emotions and the influence of attitudes and beliefs. Teachers need to know how to deal with these emotions, both the joys and the frustrations of problem solving. Traditionally, teachers have tried to control or eliminate emotional responses in mathematics classrooms, and traditional content, with its emphasis on learning low-level skills, tended to minimize the opportunities for more emotional involvement in learning. Now, however, the emphasis on higher-order thinking and problem solving makes the classroom a more exciting and more emotional place to be (Adams, this volume, Chapter 13). Research needs to provide some guidance about how to use these more intense affective responses to promote positive views of mathematics.

## Integrating Research on Cognition and Affect

In recent years, the dominant thrust in research on mathematical problem solving has been the perspective of cognitive science. Gardner (1985) notes that cognitive science has generally avoided complicating factors like affective and cultural issues in an attempt to simplify the research tasks. Such an approach is not adequate if cognitive science is to have the impact on mathematics education that its proponents would like (Schoenfeld, 1987).

A major theme of the chapters in this book is that research on affect can be incorporated into research on cognition, specifically research on mathematical problem solving. Mandler's (1984) theoretical framework for affect is particularly appropriate for analyzing the more intense emotional reactions to non-routine problems. His theory also emphasizes the role of culture, especially the importance of values, in mathematics learning (Mandler, this volume, Chapter 1); these ideas clearly have implications for the study of beliefs and attitudes about mathematics.

Mandler's ideas are particularly useful for incorporating research on affect into research on the cognitive aspects of mathematical problem solving. His emphasis on affective processes, rather than products, is characteristic of the emphasis on process in cognitive science, including research in mathematical problem solving (Schoenfeld, 1987). Although many others are now trying to incorporate affect into cognitive-processing models of learning and problem solving (Snow & Farr, 1987), and there is no intention of closing out other theoretical perspectives, Mandler's theory provides a useful guide for research that takes an integrated view of cognition and affect.

There are several areas in which cognitive approaches like Mandler's may be especially helpful in integrating research on cognition and affect. For example, substantial emphasis has been given to the general topic of metacognition in recent years, especially to the relationship between metacognition and affective factors (Lester, Garofalo, & Kroll, this volume, Chapter 6; Silver & Metzer, this volume, Chapter 5; Weinert & Kluwe, 1987). Although the relationships among beliefs, emotions, and metacognitive processes are complex, the progress that has been made is heartening. Certainly more work in this general area is appropriate, especially as it explicates the development of the kinds of positive emotions that encourage students to attempt nonroutine problems. Recent work by Malone and Lepper (1987), which proposes a cognitive theory of challenge, may help identify some new ways to encourage the positive emotions that are so crucial to the development of problem solvers. Research on intuition has also come to the fore in recent years (Fischbein, 1987; Noddings & Shore, 1984); the relationships among intuition, its cognitive and affective components, meta-cognition, and mathematical problem solving provide another challenge to researchers. All of these topics provide useful research sites for integrated investigations of both cognition and affect in mathematical problem solving.

## The Social Context of Instruction

A third theme of the chapters in this book is the important role of the social context in problem-solving instruction. When learners are asked to solve nonroutine problems, their responses are heavily influenced by their perceptions of their role and the teacher's role. As Cole and Griffin (1987) point out, the influence of contextual factors can be analyzed at several levels, from the level of the individual task to the classroom, the school, and various non-school settings. Cole and Griffin also note that these contextual factors play an especially important role in the underrepresentation of minorities and women in science and mathematics.

In mathematics education, the analyses that have been completed so far have generally been limited to the level of the individual. The role of beliefs (Schoenfeld, 1985; Silver, 1985) has received the most attention. However, the theme that recurs in many of the preceding chapters (see Cobb, Yackel, & Wood, this volume, Chapter 9, and Lester, Garofalo, & Kroll, this volume, Chapter 6) is that the influence of contextual factors is much broader than has been investigated so far. Bishop's (1988) recent work provides a comprehensive analysis of the entire notion of mathematical culture and should provide substantial assistance to investigators in this area.

Considerable progress has been made in one area of research related to mathematical culture: gender differences in mathematics learning (Fennema, this volume, Chapter 14). However, researchers have only begun to investigate the way that contextual factors influence the underrepresentation of minorities in mathematics. Recent progress in this area includes the work of Cocking and Mestre (1988) and Orr (1987), with their emphasis on language factors in learning mathematics. Much more remains to be accomplished.

Technology, as Kaput (this volume, Chapter 7) has pointed out, provides new opportunities to deal with affective factors in a changed social context. The research that is needed in the area of technology is very closely connected to the development of instructional materials. As Kaput notes, the availability of certain technologies can have a significant impact on the way that the social context can influence learning.

The technology that is available can also make possible new strategies for getting information to teachers. Tittle (1987), for example, is investigating how data on students' affective characteristics could be provided to teachers, thus making it possible for teachers to tailor instruction for students' affective as well as cognitive characteristics. If such data were available from the computers on which students were working, teachers could have summaries of data on persistence and beliefs, for example, on a regular basis. This approach could be particularly important for gender-related differences in mathematics education (Tittle, 1986), where the impact of the social context and affective factors is likely to be particularly strong.

Research on teaching provides another opportunity for learning more about how affect and the social context can influence performance in classrooms. The chapters by Thompson and Thompson (this volume, Chapter 11) and Grouws and Cramer (this volume, Chapter 10) show the pervasive influence of affect in mathematics classrooms from primary to middle-school grades. Both of these chapters focus on how expert teachers develop a classroom environment that supports mathematical problem solving. This kind of observational research is one way to find out how the "wisdom of practice" can inform our understanding of how to improve the social context for mathematics learning.

## Methods for Research on Affect and Cognition

Research on cognitive aspects of mathematical problem solving now uses a satisfying combination of qualitative and quantitative methods. Research on affective factors should develop the same range of methods as cognitive research, rather than be limited to traditional paper-and-pencil instruments. Research on beliefs, for example, can use data from interviews and observations (Lester, Garofolo, & Kroll, this volume, Chapter 6; Schoenfeld, 1985; J. Sowder, this volume, Chapter 12) as well as data from questionnaires and other more traditional instruments. The same range of methods is appropriate for research on attitudes (Marshall, this volume, Chapter 4; L. Sowder, this volume, Chapter 8).

Research on emotional aspects of mathematical problem solving also requires a range of qualitative and quantitative methods, but the quantitative methods in this case are not always appropriate for research in classrooms. In the preceding chapter, for example, Mandler (this volume, Chapter 16) indicates how he and his colleagues have measured physiological changes (including heart rate and galvanic skin response) that occur during problem solving. Although these kinds of measures are not available for researchers who work in classroom settings, there are alternative approaches.

In our current work, Cathleen Craviotto, Michele Ortega, and I have been investigating a variety of strategies for gathering reliable data on the emotions of problem solvers. So far we have been working with beginning college students who are solving nonroutine problems individually. We have audio tapes and videotapes of each problem-solving session. Although our emphasis has been on verbal reports as data (Ericsson & Simon, 1980), we are also attempting to develop methods to assess body movement, facial expression, changes in posture, and other possible indicators of emotion. Our expectation is that these physiological indicators will corroborate and expand the scope of our verbal data, which already indicate the wide range of emotions that problem solvers experience.

The verbal reports that we have emphasized so far are mainly retrospective. At the conclusion of the problem-solving episode, we have asked the problem solvers to go back over the solution to the problem and to report on what they were feeling as they worked the problem. We ask specifically about the presence of more intense emotional responses. If the students do not discuss points at which they had difficulty working the problem (i.e., points of interruption, in Mandler's sense), we ask about their feelings at these critical points in their attempts to solve the problems.

We have also done some preliminary work that involved having students go back and view the videotape with us in order to stimulate their recall of the emotions at a particular point. So far it appears that the advantage of being able to stimulate the students' recollections by replaying the videotape is offset by the disadvantage of the delay that occurs before the tape can be viewed. In contrast to this delay, it is possible to ask the students for their emotional reactions as they are trying to solve the problems (concurrent data in the sense of Ericsson & Simon, 1980). Our preliminary efforts to obtain this kind of concurrent data suggest that it is quite disruptive to ask questions about emotions during the course of a problem-solving episode.

In the case of retrospective data, we are left with the difficulty of the distortion that may occur when the students try to recall how they felt when they were solving the problem. In the case of concurrent data, we are finding that questions about emotions interfere with the students' ability to continue in the problem-solving episode. It may be possible to ameliorate the difficulties in these types of data by using more experienced subjects and by combining concurrent and retrospective data across different problem-solving episodes.

It is clear that gathering good data on the affective domain is a difficult task. Obtaining reliable data on the emotional aspects of problem solving is particularly difficult, but we believe that it is not only possible, but also necessary, to gather data on all aspects of the affective domain if we are to develop a more complete framework that can guide instruction in mathematical problem solving.

## Summary

The research reported in this book suggests that affect plays a central role in learning and instruction in mathematics. Affective factors should be incorporated in more cognitively oriented research; frequently, the additional data on affective factors could be gathered with little additional effort. However, researchers first need to become familiar with current conceptions of affect and identify theoretical frameworks that are compatible with their own research questions.

Mandler's theory (1984; this volume, Chapter 1) comes from the same cognitive tradition that guides much of research on mathematical problem solving. His theory is one that is quite compatible with the thinking of most problem-solving researchers. Describing the affective domain in terms of beliefs, attitudes, and emotions should also be compatible with most research on problem solving. Considerable attention has already been given to the role of beliefs in problem solving; now researchers need to be a bit more willing to go beyond pure cognition in order to make their investigations more applicable to real students in real classrooms.

In addition, researchers need to go beyond studies of individual problem solvers to studies of the social context of problem solving. The role of cultural factors in mathematics learning and teaching is only beginning to be part of research in mathematics education. Studies of the relationship between affect and culture should help to clarify important issues for both research and practice.

Finally, we need to refine and improve our theories of mathematics teaching and learning. The chapters by Fennema and McDonald in this book give us a good start on analyzing the strengths and weaknesses of Mandler's theory and its application to affective factors in mathematical problem solving. As research continues, more revision will be necessary. The task of revising psychological theories that are applied to mathematics learning is an interdisciplinary task. I hope that the linkages between disciplines that were established through this project will continue to flourish and contribute to the improvement of mathematics education.

*Acknowledgment.* Preparation of this paper was supported in part by the National Science Foundation (Grant No. MDR-8696142). Any opinions, conclusions, or recommendations are those of the author and do not necessarily reflect the views of the National Science Foundation.

## *References*

Abelson, R.P. (1976). Script processing in attitude formation and decision making. In J.S. Carroll & J.W. Payne (Eds.), *Cognition and social behavior* (pp. 33–45). Hillsdale, NJ: Lawrence Erlbaum Associates.

Bishop, A.J. (1988). *Mathematical enculturation: A cultural perspective on mathematics education.* Boston: Kluwer.

Brown, C.A., Carpenter, T.P., Kouba, V.L., Lindquist, M.M., Silver, E.A., & Swafford, J.O. (1988). Secondary school results for the Fourth NAEP Mathematics Assessment: Algebra, geometry, mathematical methods, and attitudes. *The Mathematics Teacher, 81,* 337–347, 397.

Brown, S.I., & Walter, M. (1983). *The art of problem posing.* Philadelphia: Franklin Institute Press.

Buxton, L. (1981). *Do you panic about maths?* London: Heinemann.

Cocking, R.R., & Mestre, J. (Eds.) (1988). *Linguistic and cultural influences on learning mathematics.* Hillsdale, NJ: Lawrence Erlbaum Associates.

Cole, M., & Griffin, P. (Eds.). (1987). *Contextual factors in education: Improving science and mathematics education for minorities and women.* Madison: Wisconsin Center for Education Research.

Commission on Standards for School Mathematics. (1987). *Curriculum and evaluation standards for school mathematics.* Reston, VA: National Council of Teachers of Mathematics.

Cooney, T.J., Grouws, D.A., & Jones, D. (1988). An agenda for research on teaching mathematics. In D.A. Grouws & T.J. Cooney (Eds.), *Effective mathematics teaching* (pp. 253–261). Reston, VA: National Council of Teachers of Mathematics. Hillsdale, NJ: Lawrence Erlbaum Associates.

D'Andrade, R.G. (1981). The cultural part of cognition. *Cognitive Science, 5,* 179–195.

Ericsson, K.A., & Simon, H.A. (1980). Verbal reports as data. *Psychological Review, 87,* 215–251.

Fennema, E., & Peterson, P. (1985). Autonomous learning behavior: A possible explanation of gender-related differences in mathematics. In L.C. Wilkinson & C. Marrett (Eds.), *Gender influences in classroom interaction* (pp. 17–35). Orlando, FL: Academic Press.

Fennema, E., & Sherman, J.A. (1976). Fennema-Sherman Mathematics Attitude Scales: Instruments designed to measure attitudes toward the learning of mathematics by females and males. *Journal for Research in Mathematics Education, 7,* 324–326.

Fischbein, E. (1987). *Intuition in science and mathematics.* Boston: Reidel.

Gardner, H. (1985). *The mind's new science.* New York: Basic Books.

Gentry, W.M., & Underhill, R. (1987). A comparison of two palliative methods of intervention for the treatment of mathematics anxiety among female college students. In J.C. Bergeron, N. Herscovics, & C. Kieran (Eds.), *Proceedings of the Eleventh International Conference on the Psychology of Mathematics Education* (Vol. 1, pp. 99–105). Montreal: University of Montreal.

Haladyna, T., Shaughnessy, J., & Shaughnessy, J.M. (1983). A causal analysis of attitude toward mathematics. *Journal for Research in Mathematics Education, 14,* 19–29.

Kulm, G. (1980). Research on mathematics attitude. In R.J. Shumway (Ed.), *Research in mathematics education* (pp. 356–387). Reston, VA: National Council of Teachers of Mathematics.

Lave, J., Murtaugh, M., & de la Rocha, O. (1984). The dialectic of arithmetic in grocery shopping. In B. Rogoff & J. Lave (Eds.), *Everyday cognition: Its development in social context* (pp. 67–94). Cambridge, MA: Harvard University Press.

Lawler, R.W. (1981). The progressive construction of mind. *Cognitive Science, 5,* 1–30.

Leder, G.C. (1987). Attitudes towards mathematics. In T.A. Romberg & D.M. Stewart (Eds.), *The monitoring of school mathematics* (pp. 261–277). Madison: Wisconsin Center for Education Research.

Malone, T.W., & Lepper, M.R. (1987). Making learning fun: A taxonomy of intrinsic motivations for learning. In R.E. Snow & M.J. Farr (Eds.), *Aptitude, learning, and instruction: Vol. 3. Conative and affective process analyses* (pp. 223–253). Hillsdale, NJ: Lawrence Erlbaum Associates.

Mandler, G. (1984). *Mind and body: Psychology of emotion and stress.* New York: Norton.

Meyer, M.R., & Fennema, E. (1988). Girls, boys, and mathematics. In T.R. Post (Ed.), *Teaching mathematics in grades K–8: Research-based methods* (pp. 406–425). Boston: Allyn & Bacon.

Noddings, N., & Shore, P.J. (1984). *Awakening the inner eye: Intuition in education.* New York: Teachers College Press.

Orr, E.W. (1987). *Twice as less: Black English and the performance of black students in mathematics and science.* New York: Norton.

Peterson, P.L., & Barger, S.A. (1985). Attribution theory and teacher expectancy. In J.B. Dusek (Ed.), *Teacher expectancies* (pp. 159–184). Hillsdale, NJ: Lawrence Erlbaum Associates.

Prawat, R.S., Byers, J.L., & Anderson, A.H. (1983). An attributional analysis of teachers' affective reactions to student success and failure. *American Educational Research Journal, 20,* 137–152.

Reyes, L.H. (1984). Affective variables and mathematics education. *Elementary School Journal, 84,* 558–581.

Schoenfeld, A.H. (1985). *Mathematical problem solving.* Orlando, FL: Academic Press.

Schoenfeld, A.H. (1987). Cognitive science and mathematics education: An overview. In A.H. Schoenfeld (Ed.), *Cognitive science and mathematics education* (pp. 1–31). Hillsdale, NJ: Lawrence Erlbaum Associates.

Silver, E.A. (1985). Research on teaching mathematical problem solving: Some underrepresented themes and needed directions. In E.A. Silver (Ed.), *Teaching and learning mathematical problem solving: Multiple research perspectives* (pp. 247–266). Hillsdale, NJ: Lawrence Erlbaum Associates.

Skemp, R.R. (1979). *Intelligence, learning, and action.* New York: Wiley.

Snow, R.E., & Farr, M.J. (Eds.). (1987). *Aptitude, learning, and instruction: Vol. 3. Conative and affective process analyses.* Hillsdale, NJ: Lawrence Erlbaum Associates.

Thompson, A.G. (1984). The relationship of teachers' conceptions of mathematics and mathematics teaching to instructional practice. *Educational Studies of Mathematics, 15,* 105–127.

Tittle, C.K. (1986). Gender research and education. *American Psychologist, 41,* 1161–1168.

Tittle, C.K. (1987). *A project to improve mathematics instruction for women and minorities: Comprehensive assessment and mathematics.* Unpublished manuscript, City University of New York, Graduate School and University Center, New York.

von Glasersfeld, E. (1987). Learning as a constructive activity. In C. Janvier (Ed.), *Problems of representation in the teaching and learning of mathematics* (pp. 3–17). Hillsdale, NJ: Lawrence Erlbaum Associates.

Wagner, S., Rachlin, S.L., & Jensen, R.J. (1984). *Algebra learning project: Final Report.* Athens: University of Georgia.

Weiner, B. (1986). *An attributional theory of motivation and emotion.* New York: Springer-Verlag.

Weinert, F.E., & Kluwe, R.H. (Eds.). (1987). *Metacognition, motivation, and understanding.* Hillsdale, NJ: Lawrence Erlbaum Associates.

# Author Index

# Subject Index